Solutions Manual

Introduction to Analysis and Design of Equilibrium Staged Separation Processes

Solutions Manual

Introduction to Analysis and Design of Equilibrium Staged Separation Processes

RAJINDER PAL

Professor of Chemical Engineering
University of Waterloo
Ontario, Canada

CWP

Central West Publishing

This edition has been published by Central West Publishing, Australia

For more information about the books published by Central West Publishing, please visit https://centralwestpublishing.com

Disclaimer

Every effort has been made by the publisher and author while preparing this book; however, no warranties are made regarding the accuracy and completeness of the content. The publisher and author disclaim without any limitation all warranties as well as any implied warranties about sales, along with fitness of the content for a particular purpose. Citation of any website and other information sources does not mean any endorsement from the publisher and author. For ascertaining the suitability of the contents contained herein for a particular lab or commercial use, consultation with the subject expert is needed. In addition, while using the information and methods contained herein, the practitioners and researchers need to be mindful for their own safety, along with the safety of others, including the professional parties and premises for which they have professional responsibility. To the fullest extent of law, the publisher and author are not liable in all circumstances (special, incidental, and consequential) for any injury and/or damage to persons and property, along with any potential loss of profit and other commercial damages due to the use of any methods, products, guidelines, procedures contained in the material herein.

NATIONAL LIBRARY OF AUSTRALIA

A catalogue record for this book is available from the National Library of Australia

ISBN (print): 978-1-925823-77-6

Contents

Preface

This Solutions Manual gives complete solutions of all the practice problems given at the end of each chapter (total of 16 chapters) of the text INTRODUCTION TO ANALYSIS AND DESIGN OF EQUILIBRIUM STAGED SEPARATION PROCESSES. For the convenience of the readers, the practice problems given in the text have been restated before providing the solution. Some of the typos detected in the text have also been corrected.

Waterloo, December 2019 *Rajinder Pal*

Author

Rajinder Pal is a Professor in the Department of Chemical Engineering at the University of Waterloo since 1991, and a Fellow of the Chemical Institute of Canada. He received degrees in Chemical Engineering from I.I.T. Kanpur (B.Tech. 1981) and the University of Waterloo (Ph.D. 1987), followed by a post-doctoral fellowship at the University of Alberta. His research interests include complex fluid rheology, interfacial and transport phenomena, physical properties of composite materials, and non-equilibrium thermodynamics. He has published close to 150 refereed journal articles, 2 books, and 5 book chapters on these and related subjects, including 6 papers in engineering education journals. He has taught over ten different undergraduate and graduate courses related to equilibrium staged separation processes, fluid mechanics, heat and mass transfer, physical chemistry and thermodynamics.

Professor Pal has been the recipient of several awards including the Sandford Fleming Foundation Teaching Excellence Award (twice), an Outstanding Performance Award from the University of Waterloo, and the Syncrude Canada Innovation Award from the Canadian Society for Chemical Engineering. He serves on several journal editorial boards, and was an Associate Editor of the Canadian Journal of Chemical Engineering for twelve years. He is a registered Professional Engineer in the province of Ontario, Canada.

Chapter 1

Introduction to Equilibrium Stage Operations

Problem 1

Plot the variations of partial pressures and total pressure for a binary vapor-liquid system as functions of mole fraction of a more volatile species A (at a fixed temperature). Both vapor mixture and liquid solution are ideal in nature, that is, the Raoult law is applicable over the entire composition range. The vapor pressures of the pure species at the given temperature are as follows: $P_A^{Sat} = 135.5\ kPa$ and $P_B^{Sat} = 54\ kPa$

Solution

The vapor pressures of pure species are:

$$P_A^{Sat} = 135.5\ kPa$$

$$P_B^{Sat} = 54\ kPa$$

From Raoult's law, the partial pressure of species A is:

$$P_A = x_A P_A^{Sat} = 135.5 x_A$$

The plot of partial pressure P_A as a function of mole fraction of a more volatile species x_A is shown below. It is a linear plot passing through the origin and has a slope of 135.5 kPa.

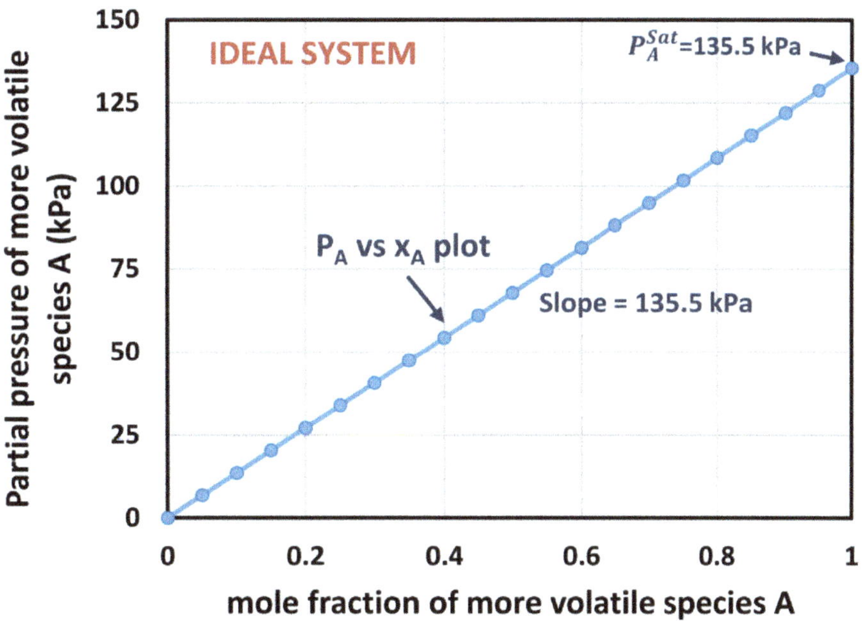

From Raoult's law, the partial pressure of species B is:

$$P_B = x_B P_B^{Sat} = 54(1 - x_A)$$

The plot of partial pressure P_B as a function of mole fraction of a more volatile species x_A is shown below. It is a linear plot with y-intercept of 54 kPa and a slope of -54 kPa.

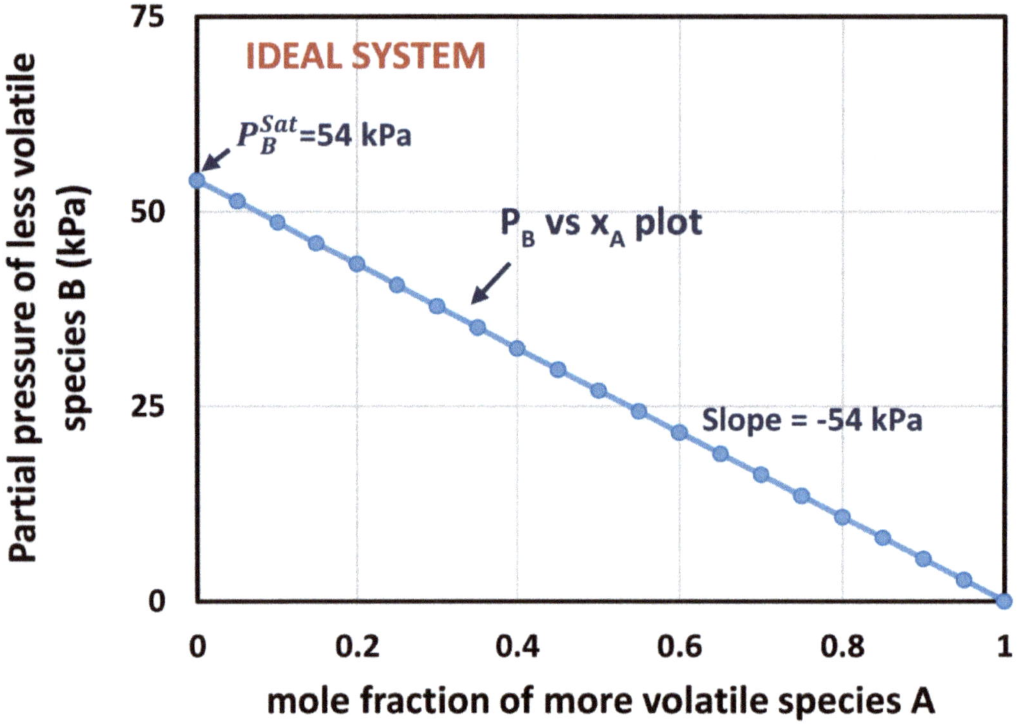

From Raoult's law, the total pressure of a binary mixture is:

$$P = x_A P_A^{Sat} + x_B P_B^{Sat} = x_A P_A^{Sat} + (1 - x_A) P_B^{Sat} = x_A \left(P_A^{Sat} - P_B^{Sat}\right) + P_B^{Sat}$$

Upon substitution of the values of $P_A^{Sat} = 13.5.5\ kPa$ and $P_B^{Sat} = 54\ kPa$, we get:

$$P = 54 + 81.5 x_A$$

The plot of total pressure P as a function of mole fraction of a more volatile species x_A is shown below. It is linear plot with a slope of 81.5 kPa. Note that when $x_A = 0$, the total pressure is equal to $P_B^{Sat} = 54\ kPa$ as the liquid and vapor are pure species B. Similarly when $x_A = 1.0$, the total pressure is equal to $P_A^{Sat} = 135.5\ kPa$ as the liquid and vapor are pure species A.

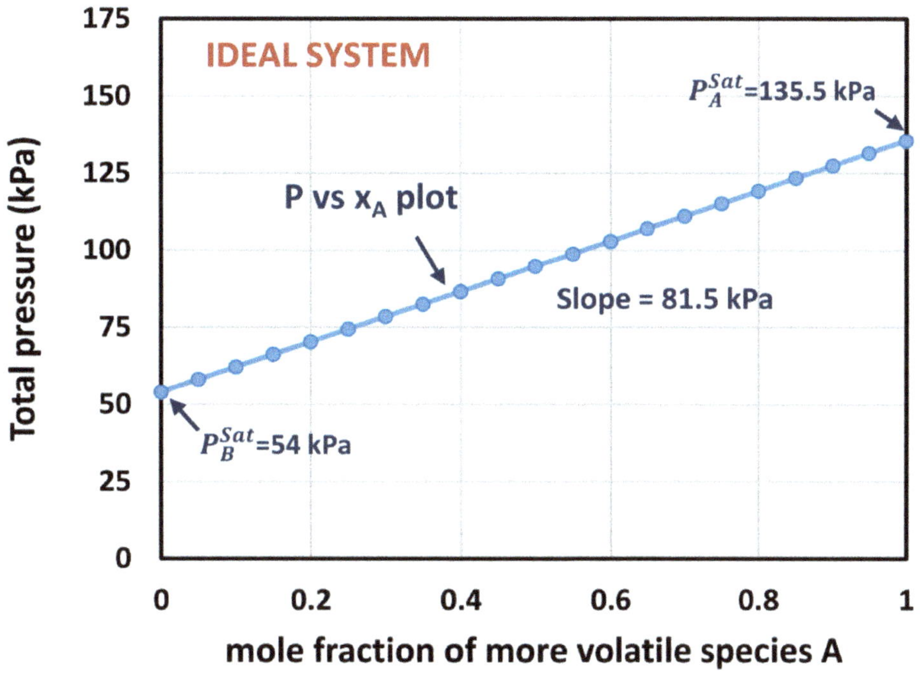

Problem 2

The solubility data of nitrogen gas in liquid water at 25 °C is given as follows:

Mole fraction of nitrogen in water, x_A	**Partial pressure of nitrogen, P_A (bars)**
0	0
0.00002	1.728
0.00004	3.456
0.00005	4.32
0.00007	6.048
0.0001	8.64
0.00015	12.96

Plot the data as P_A $versus$ x_A. From the slope of the plot, determine the Henry law constant K_H.

Solution

The data are plotted below. As expected, the data follow a linear relationship (Henry's law) passing through origin and having a slope $K_H = 86400\ bars$.

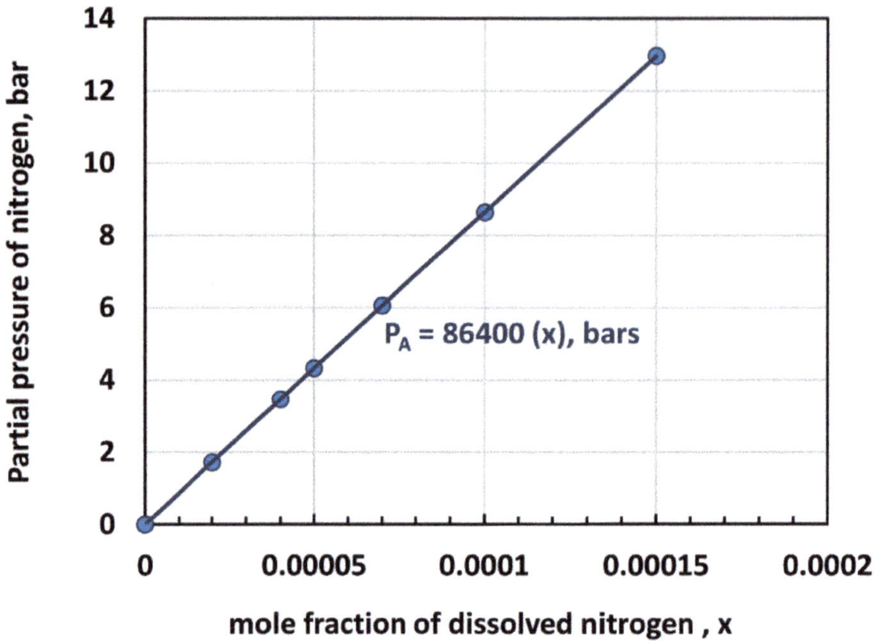

Problem 3

At 298 K, the Henry law constant of oxygen dissolved in water is 4.259×10^4 atm. Estimate the solubility of oxygen in moles per liter in water at 298 K and oxygen partial pressure of 5 atm. Given that the molecular weight of oxygen is 16, determine the solubility of oxygen in water in grams/litre. The density of water is 997 g/litre and its molecular weight is 18.02 g/mole.

Solution

Using Henry's law:

$$P_{O_2} = (4.259 \times 10^4)x$$

$$x = \frac{P_{O_2}}{4.259 \times 10^4} = \frac{5}{4.259 \times 10^4} = 1.174 \times 10^{-4}$$

$$x = \frac{n_{O_2}}{n_{O_2} + n_W}$$

where n is the number of moles of species in the liquid solution. Upon re-arrangement of this equation, we obtain

$$n_{O_2} = \frac{x}{1-x} n_W$$

As the mole fraction x is small, we can calculate the number of moles of water per litre of solution as:

$$n_W = \frac{\rho_W}{M_W}$$

where ρ_W is the density of water and M_W is the molecular weight of water. Thus

$$n_W = \frac{\rho_W}{M_W} = \frac{997}{18.02} = 55.33\ moles\ /litre$$

Therefore, the solubility of oxygen in moles per liter is

$$n_{O_2} = \frac{x}{1-x} n_W = \frac{1.174 \times 10^{-4}}{1 - 1.174 \times 10^{-4}}(55.33) = 6.497 \times 10^{-3}\ moles/litre$$

Since the molecular weight of oxygen is 16, we can convert this molar concentration to mass concentration as:

$$m_{O_2} = 6.497 \times 10^{-3} \times 16 = 0.104\ g/litre$$

Chapter 2

Single-Stage Gas Absorption

Problem 1

A gas mixture containing 98.5 mole % air and 1.5 mole% acetone is passed through an equilibrium stage so that acetone can be absorbed in water stream. The molar flow rate of gas mixture is 25 kmol/h. The pure water stream enters at a molar flow rate of 100 kmol/h. The Henry law is $y = 2.5x$. Determine the mole fractions of acetone in the exit liquid and gas streams. What is the percent recovery of acetone in the gas phase? You can assume that the molar flow rates of gaseous and liquid streams do not change in passing through the equilibrium stage.

Solution

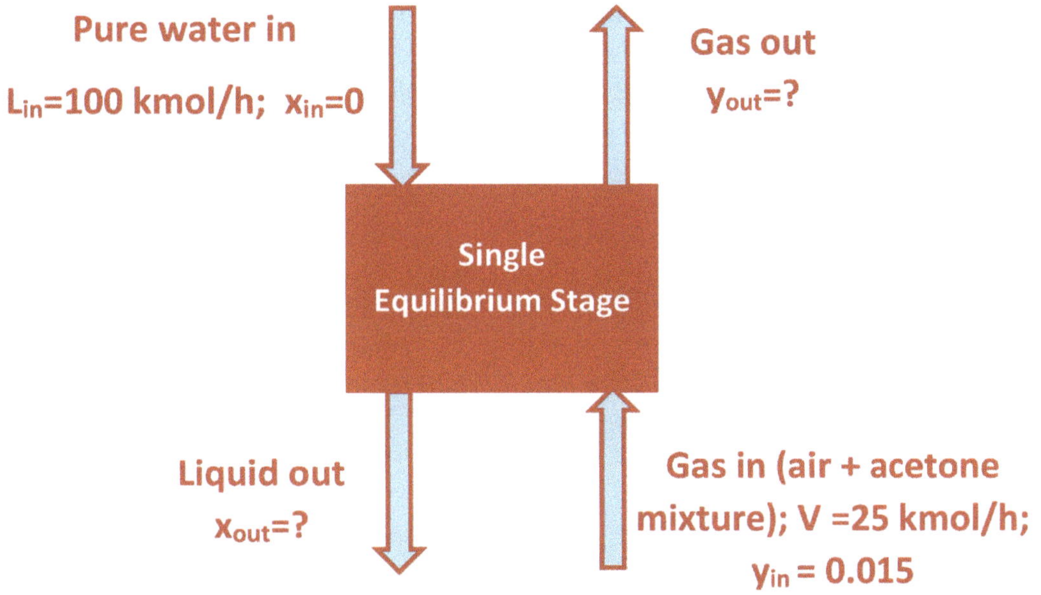

We assume $L_{in} = L_{out} = L = 100\ kmol/h$ and $V_{in} = V_{out} = V = 25\ kmol/h$. The equilibrium constant K is 2.5. The absorption factor A can be calculated as:

$$A = \frac{L/V}{K} = \frac{100/25}{2.5} = 1.6$$

As shown in the diagram, $x_{in} = 0$. Consequently,

$$\frac{y_{out}}{y_{in}} = \left(\frac{1}{1+A}\right) = \frac{1}{1+1.6} = 0.3846$$

$$y_{out} = 0.3846 \times 0.015 = 5.77 \times 10^{-3}$$

$$x_{out} = y_{out}/K = 5.77 \times 10^{-3}/(2.5) = 2.31 \times 10^{-3}$$

The percent recovery of solute from the gas phase to the liquid phase can be calculated as follows:

$$\%\ Recovery\ of\ Solute = 100\left(1 - \frac{y_{out}}{y_{in}}\right) = 100\left(1 - \frac{5.77\times10^{-3}}{0.015}\right) = 61.54$$

The recovery of solute is good, 61.54 % , in this single-stage absorption process.

Note that the absorption factor A is large in the present case $(A = 1.6)$ and therefore, the percent recovery of solute is good.

Problem 2

A gas mixture consisting of air and sulphur dioxide (SO_2) is contacted in a single-stage contactor with pure water at 2 bar and 295 K. The partial pressure of SO_2 in the original gas is 0.20 bar. The inlet gas molar flow rate is 6 kmol/h. The inlet pure water stream has a molar flow rate 2 kmol/h. The exit gaseous and liquid streams are at equilibrium with each other. The Henry law is $Y = 15X$ where Y is the mole ratio of solute in the gas phase and X is the mole ratio of solute in the liquid phase. The mole ratios are defined as: $Y = y/(1-y)$ and $X = x/(1-x)$. Determine the molar flow rates of exit gaseous and liquid streams. What is the percent recovery of solute in the gas phase? Justify the estimated percent recovery in terms of the absorption factor. The gas and liquid molar rates should not be assumed constant in the present case.

Solution

Refer to the diagram given below

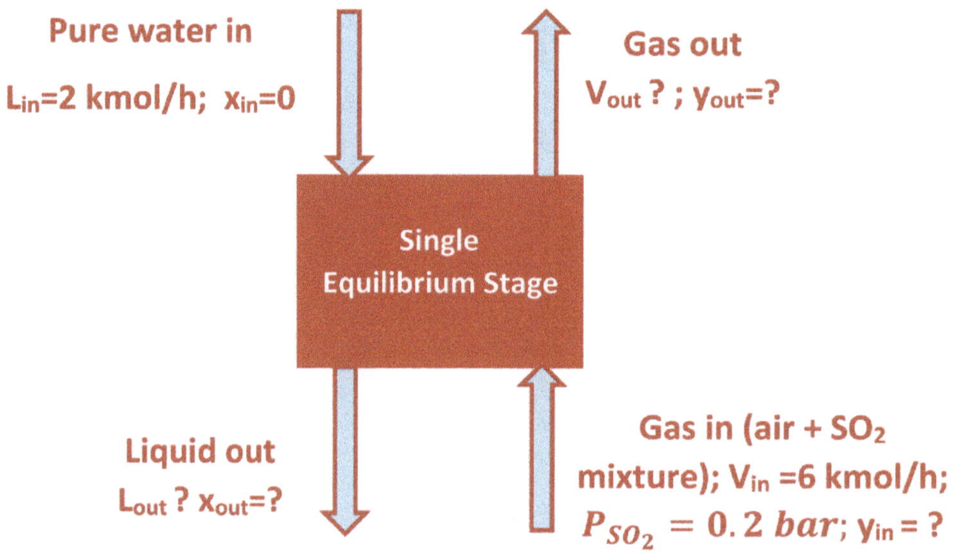

Molar flow rates of exit gaseous and liquid streams?

$$L' = L_{in}(1 - x_{in}) = 2(1 - 0) = 2\ kmol/h$$

$$P_{SO_2} = y_{in}P\ ;\ y_{in} = P_{SO_2}/P = 0.2/2 = 0.1$$

$$V' = V_{in}(1 - y_{in}) = 6(1 - 0.1) = 5.4\ kmol/h$$

$$Y_{in} = \frac{y_{in}}{1 - y_{in}} = \frac{0.10}{1 - 0.10} = 0.1111$$

$$X_{in} = \frac{x_{in}}{1 - x_{in}} = \frac{0}{1 - 0} = 0$$

$$\frac{L'}{V'} = \frac{2}{5.4} = 0.3704$$

The absorption factor A' is:

$$A' = \frac{L'/V'}{K'} = \frac{2/5.4}{15} = 0.0247$$

As $X_{in} = 0$,

$$\frac{Y_{out}}{Y_{in}} = \left(\frac{1}{1 + A'}\right) = \frac{1}{1 + 0.0247} = 0.9759$$

$$Y_{out} = 0.9759 \times 0.1111 = 0.1084$$

The equilibrium relation gives:

$$Y_{out} = 15(X_{out})$$

Thus:

$$X_{out} = Y_{out}/15 = 0.1084/15 = 7.22 \times 10^{-3}$$

The mole fractions can be calculated from the mole ratios as follows:

$$y_{out} = \frac{Y_{out}}{1 + Y_{out}} = \frac{0.1084}{1 + 0.1084} = 0.0978$$

$$x_{out} = \frac{X_{out}}{1 + X_{out}} = \frac{7.22 \times 10^{-3}}{1 + 7.22 \times 10^{-3}} = 7.18 \times 10^{-3}$$

The molar flow rates of the exit gas and liquid streams can now be calculated:

$$L_{out} = \frac{L'}{1 - x_{out}} = \frac{2}{1 - 7.18 \times 10^{-3}} = 2.015\ kmol/h$$

Thus L has increased from 2.0 to 2.015 kmol/h.

$$V_{out} = \frac{V'}{1 - y_{out}} = \frac{5.4}{1 - 0.0978} = 5.985\ \ kmol/h$$

The gas flow rate has decreased from 6.0 to 5.985 kmol/h.

The percent recovery of solute?

$$\%\ Recovery\ of\ Solute = 100\left(1 - \frac{Y_{out}}{Y_{in}}\right) = 100\left(1 - \frac{0.1084}{0.1111}\right) = 2.43$$

The recovery of solute of 2.43% is small in this single-stage absorption process. This low recovery in a single stage absorption process is not surprising as the absorption factor $A' = 0.0247$ is small due to low solubility of solute in the solvent as indicated by a large value of Henry's law constant.

Problem 3

A solute is to be recovered into a liquid solvent from an inert carrier gas using single-stage absorption. The gas enters the absorber at a total molar flow rate of 350 kmol/h with $y = 0.35$ and leaves the absorber with $y = 0.05$. Pure solvent enters the absorber at a molar flow rate of 1200 kmol/h. The carrier gas is insoluble in the solvent and the solvent is nonvolatile.

a) Estimate the mole fraction of solute in the liquid stream leaving the absorber.
b) Calculate the total molar flow rates of the exit streams

Solution

a) The mole fraction of solute in the exit liquid stream ?

Refer to the diagram given below

$$L' = L_{in}(1 - x_{in}) = 1200(1 - 0) = 1200\ kmol/h$$

$$V' = V_{in}(1 - y_{in}) = 350(1 - 0.35) = 227.5\ kmol/h$$

$$Y_{in} = \frac{y_{in}}{1 - y_{in}} = \frac{0.35}{1 - 0.35} = 0.5385$$

$$X_{in} = \frac{x_{in}}{1 - x_{in}} = \frac{0}{1 - 0} = 0$$

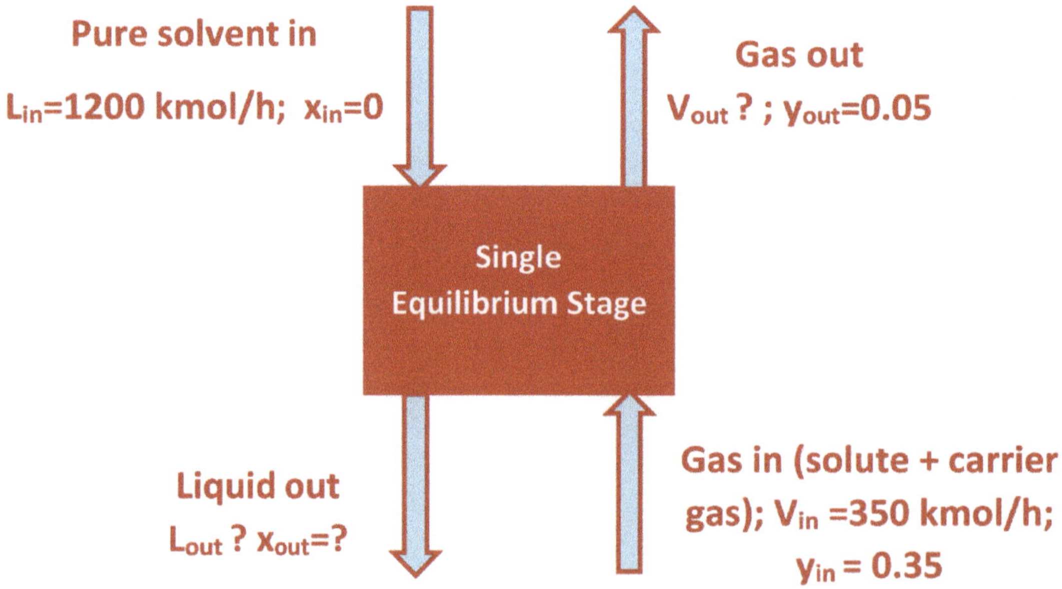

$$Y_{out} = \frac{y_{out}}{1 - y_{out}} = \frac{0.05}{1 - 0.05} = 0.0526$$

$$\frac{L'}{V'} = \frac{1200}{227.5} = 5.275$$

Equation of the operating line:

$$Y_{out} = Y_{in} - \left(\frac{L'}{V'}\right)(X_{out} - X_{in}) = 0.5385 - 5.275X_{out}$$

Given $Y_{out} = 0.0526$, we can calculate X_{out} from the above operating line:

$$X_{out} = 0.092$$

The mole fraction of solute in the exit liquid stream can be calculated from the mole ratio as follows:

$$x_{out} = \frac{X_{out}}{1 + X_{out}} = \frac{0.092}{1 - 0.092} = 0.0842$$

b) The total molar flow rates of the exit streams?

$$L_{out} = \frac{L'}{1 - x_{out}} = \frac{1200}{1 - 0.0842} = 1310.33\ kmol/h$$

Thus L has increased from 1200 to 1310.33 kmol/h.

$$V_{out} = \frac{V'}{1 - y_{out}} = \frac{227.5}{1 - 0.05} = 239.47\ kmol/h$$

The gas flow rate has decreased from 350 to 239.47 kmol/h.

Problem 4

A solute is to be recovered into a liquid solvent from an inert carrier gas using single-stage absorption. The gas enters the absorber at a total molar flow rate of 350 kmol/h with $y = 0.35$. Pure solvent enters the absorber at a molar flow rate of 1200 kmol/h. The carrier gas is insoluble in the solvent and the solvent is nonvolatile.

Construct the following plots: (a) y-x plot for the operating line, covering x range of 0 to 0.1; (b) Y-X plot for the operating line, covering x range of 0 to 0.1.

Which coordinates (y-x or Y-X) give a *linear* operating line?

Solution

Refer to the diagram given as

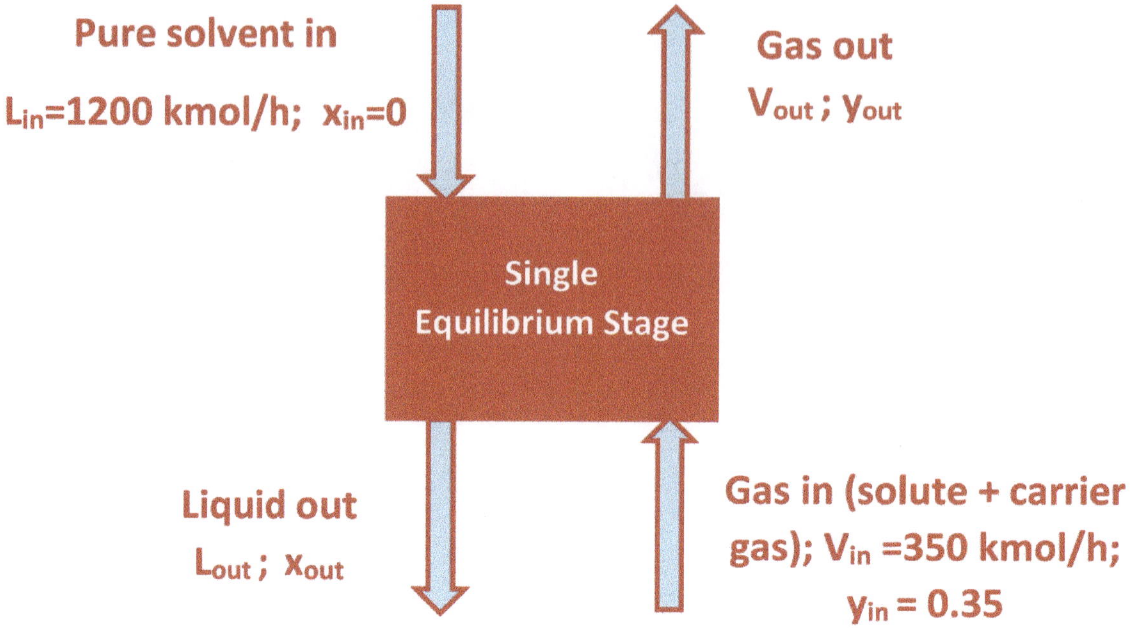

Equation of the operating line:

$$Y = Y_{in} - \left(\frac{L'}{V'}\right)(X - X_{in})$$

$$Y_{in} = \frac{y_{in}}{1 - y_{in}} = \frac{0.35}{1 - 0.35} = 0.5385$$

$$X_{in} = \frac{x_{in}}{1 - x_{in}} = \frac{0}{1 - 0} = 0$$

$$L' = L_{in}(1 - x_{in}) = 1200(1 - 0) = 1200\ kmol/h$$

$$V' = V_{in}(1 - y_{in}) = 350(1 - 0.35) = 227.5\ kmol/h$$

$$\frac{L'}{V'} = \frac{1200}{227.5} = 5.2747$$

Thus, the operating line on Y-X co-ordinates is:

$$Y = 0.5385 - 5.2747(X)$$

The operating line on y-x co-ordinates is:

$$\frac{y}{1 - y} = 0.5385 - 5.2747\left(\frac{x}{1 - x}\right)$$

Note that the operating line relates the compositions of the exit streams.

(a) The plot of the operating line on y-x (mole fractions) co-ordinates is shown in the figure below. It is a curve passing through the point (x_{in}, y_{in}), that is, (0, 0.35).

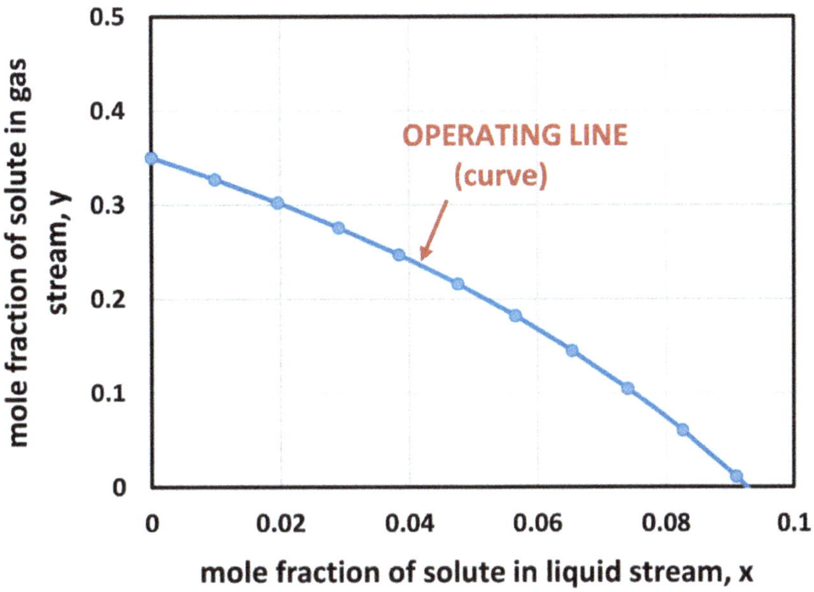

(b) The plot of the operating line on Y-X (mole ratios) co-ordinates is shown in the figure below. It is a straight line passing through the point (X_{in}, Y_{in}), that is, (0, 0.5385). It has a slope of $-(L'/V') = -5.275$.

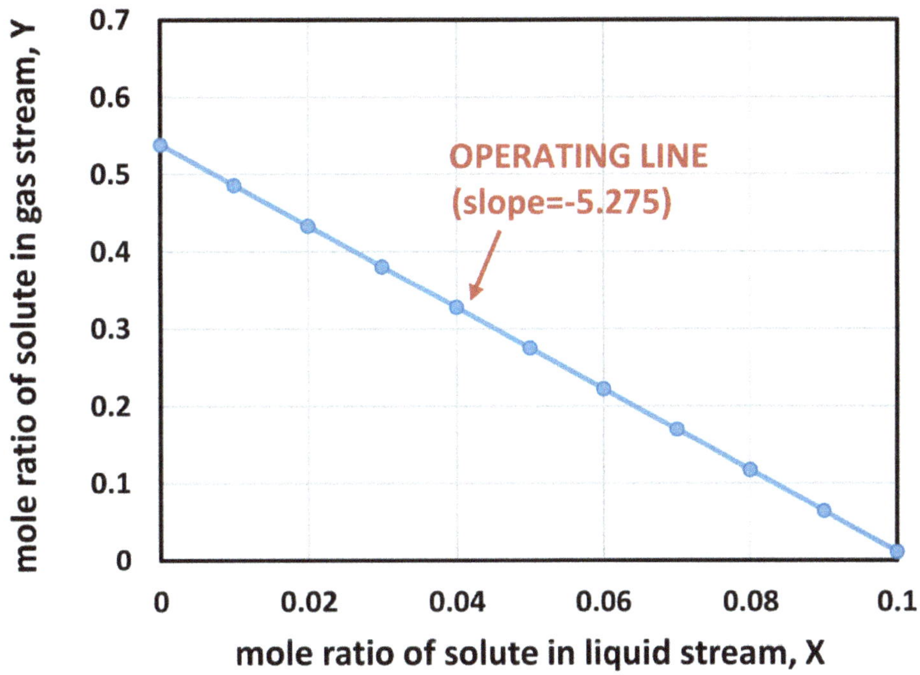

Thus, the advantage of using mole ratios for concentrated systems is that the operating line is a straight line. The mole fraction coordinates give a curve for the operating line.

Chapter 3

Single-Stage Liquid Stripping

Problem 1

A hydrocarbon mixture consisting of 5 mole% propane and 95 mole% non-volatile hydrocarbon is being stripped by pure superheated steam using a single stage process. The ratio of steam molar flow rate to hydrocarbon molar flow rate is 1/10, that is, $V/L = 0.1$. The equilibrium relation is $y = 30x$, where y and x are mole fractions of propane in steam and hydrocarbon mixture, respectively.

a) Determine the mole fractions of propane in the exit streams.
b) What is the value of stripping factor?
c) Calculate percent removal of propane from the hydrocarbon mixture

You can assume steam is an inert gas and it does not condense.

Solution

Refer to the diagram given below

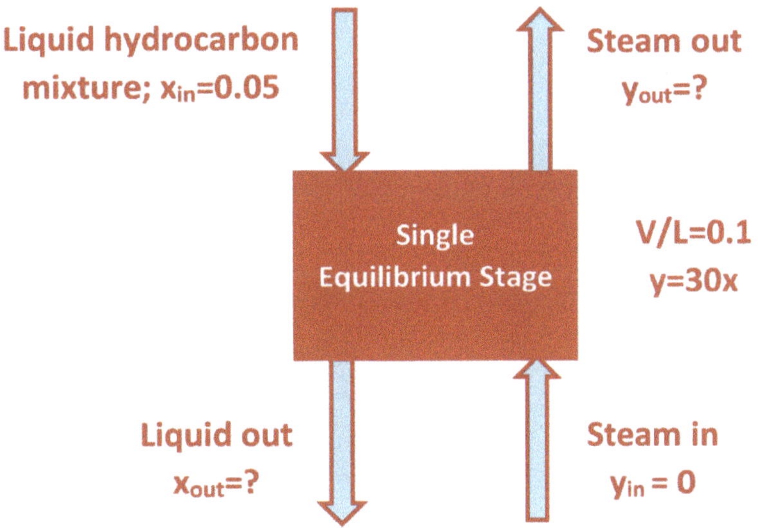

a) As $y_{in} = 0$, we can calculate the mole fraction in the exit liquid stream from:

$$\frac{x_{out}}{x_{in}} = \left(\frac{1}{1+S}\right)$$

The stripping factor is:

$$S = \frac{K}{L/V} = 30 \times 0.10 = 3.0$$

Thus,

$$\frac{x_{out}}{x_{in}} = \left(\frac{1}{1+3}\right) = 0.25$$

$$x_{out} = 0.25 \times 0.05 = 0.0125$$

$$y_{out} = 30x_{out} = 30 \times 0.0125 = 0.375$$

b) The value of the stripping factor is calculated in part (a) as 3.0

c) The percent removal of propane from the hydrocarbon mixture is as follows:

$$\%\ Removal\ of\ Solute = 100\left(1 - \frac{x_{out}}{x_{in}}\right) = 100(1 - 0.25) = 75\%$$

Problem 2

A single-stage stripping process is being used to remove ammonia from wastewater using pure air as the stripping gas. The water entering the process contains 0.05 mole fraction ammonia. The ammonia mole fraction in the exit water stream is 0.01. The equilibrium relation is $y = 1.5x$, where y and x are mole fractions of ammonia in air and water streams at equilibrium, respectively. Determine the ratio V/L. You can assume that the molar flow rates of gas and liquid streams are constant.

Solution

Refer to the diagram given below

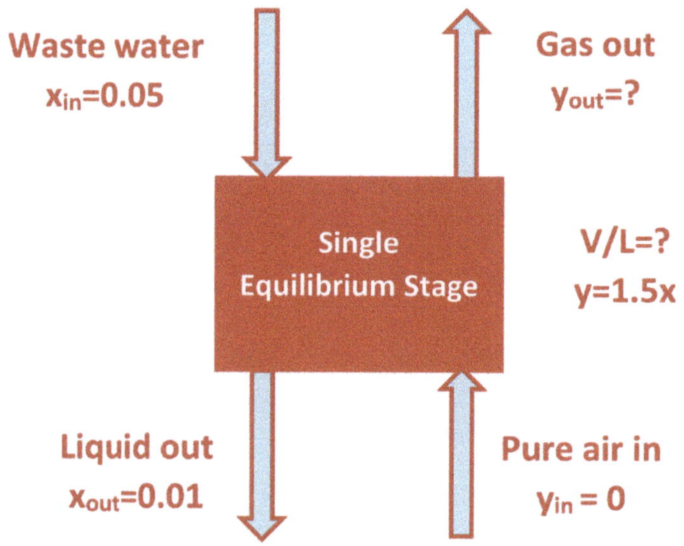

As $y_{in} = 0$, we can calculate the stripping factor from:

$$\frac{x_{out}}{x_{in}} = \left(\frac{1}{1+S}\right) = \frac{0.01}{0.05} = 0.2$$

The stripping factor calculated from above relation is 4. Thus,

$$S = \frac{K}{L/V} = 4.0$$

Since $K = 1.5$,

$$L/V = 1.5/4 = 0.375$$

Therefore,

$$V/L = 2.67$$

Chapter 4

Multi-Stage Gas Absorption

Problem 1

Acetone is being absorbed from acetone-air mixture in a multi-stage absorption tower using pure water as a solvent. The gas entering the tower contains 1 mole percent acetone and 80 percent of it is to be absorbed. The molar flow rates of gas and water entering the tower are 25 kmol/h and 85 kmol/h, respectively. The process operates isothermally (300 K) at a total pressure of 101.3 kPa. Assume that the equilibrium relationship for this system at the temperature and pressure of the process is given as: $y = 2.5x$, where y and x are mole fractions of acetone in gas and liquid phases, respectively at equilibrium.

a) Determine the number of equilibrium stages required using the McCabe-Thiele graphical method.

b) Determine the number of stages using the Kremser equation

You can assume that the molar flow rates of gas and liquid streams are constant as the system is dilute in solute.

Solution

a) Refer to the column schematic shown below.

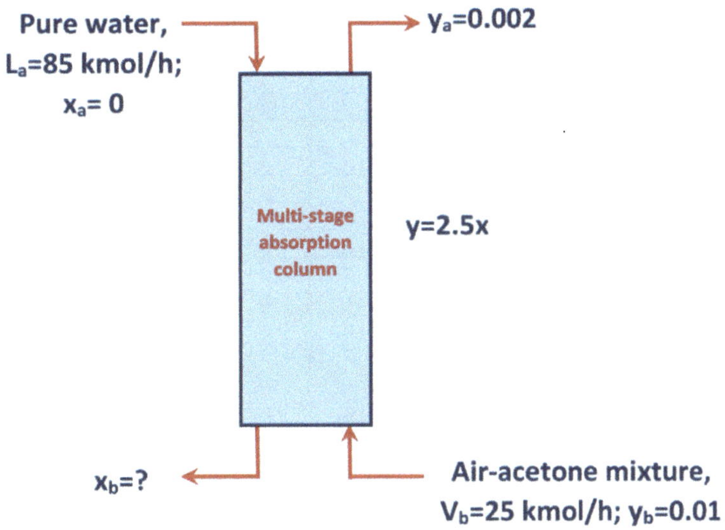

As 80% of entering acetone is removed, the mole fraction of acetone in the gas leaving the column is as follows:

$$y_a = 0.01 - (0.80 \times 0.01) = 0.002$$

The mole fraction of acetone in the liquid leaving the column at the bottom, that is, x_b , can be calculated from the operating line using the terminal points:

$$y_b = y_a + (\frac{L}{V})(x_b - x_a)$$

$$0.01 = 0.002 + (85/25)(x_b - 0)$$

$$x_b = 0.002353$$

Thus the operating line passes through the points (0, 0.002) and (0.002353, 0.01). The following figure shows the plots of the operating and equilibrium lines. The stepping off the stages using the McCabe-Thiele method is also shown in the figure. The number of equilibrium stages required is 2 full stages and one partial stage ~1/3.

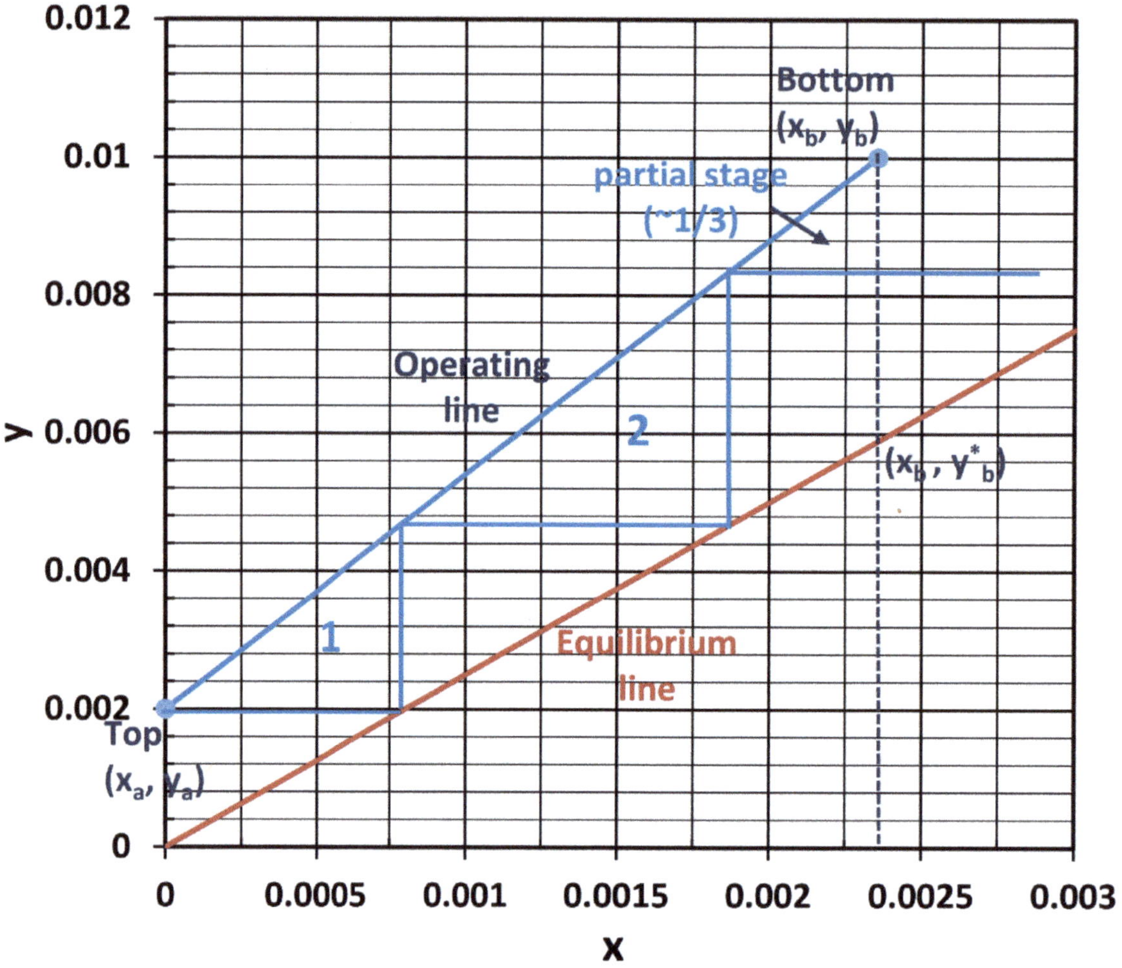

b) Let us now apply the Kremser equation to estimate the number of equilibrium stages. The absorption factor is

$$A = (L/V)/K = (85/25)/2.5 = 1.36$$

As $A \neq 1$, the following form of Kremser equation is applicable:

$$N = \frac{ln\left(\frac{y_b - y_b^*}{y_a - y_a^*}\right)}{ln(A)} \qquad when\ A \neq 1$$

$$y_b = 0.01, \;\; y_b^* = Kx_b = 2.5 \times 0.002353 = 0.005882$$

$$y_a = 0.002, \;\; y_a^* = Kx_a = 2.5 \times 0 = 0$$

Thus,

$$N = \frac{ln\left(\frac{y_b - y_b^*}{y_a - y_a^*}\right)}{ln(A)} = \frac{ln\left(\frac{0.01-0.005882}{0.002-0}\right)}{\ln(1.36)} = 2.35$$

This value is almost the same as that obtained graphically using the McCabe-Thiele method.

Problem 2

A multi-stage countercurrent absorption tower is absorbing ethyl alcohol from an inert gas stream containing 2 mole percent ethyl alcohol. Pure water is being used as a solvent. The process operates isothermally (303 K) at a total pressure of 101.3 kPa. The molar flow rate of gas stream is 90 kmol/h. It is desired to remove 85% of the alcohol. Assume that the equilibrium relationship for this system at the temperature and pressure of the process is given as: $y = 0.7x$, where y and x are mole fractions of ethyl alcohol in gas and liquid phases, respectively at equilibrium. The actual liquid molar flow rate is 1.6 times the minimum liquid molar flow rate.

a) Determine the number of equilibrium stages required using the McCabe-Thiele graphical method.

b) Determine the number of stages using the Kremser equation

You can assume that the molar flow rates of gas and liquid streams are constant as the system is dilute in solute.

Solution

a) Refer to the column schematic shown below. As 85% of entering ethyl alcohol is removed, the mole fraction of ethyl alcohol in the gas leaving the column is as follows:

$$y_a = 0.02 - (0.85 \times 0.02) = 0.003$$

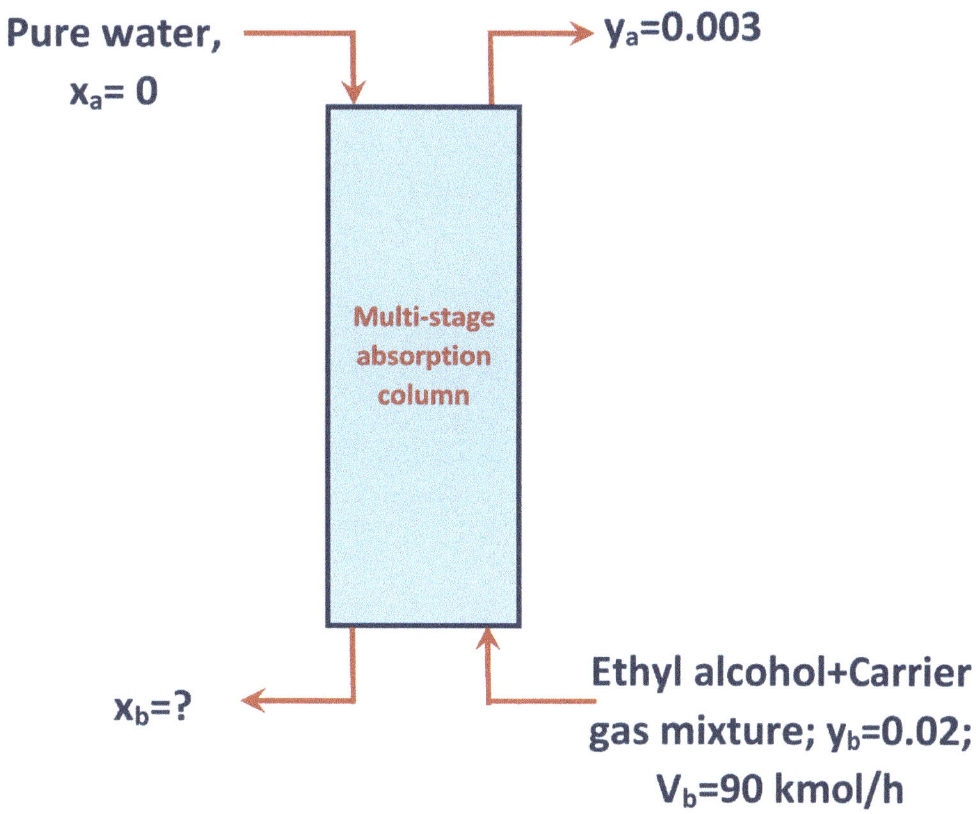

The minimum slope of the operating line is when the operating line just touches the equilibrium line, as shown in the figure below. Thus,

$$Minimum\ slope = (L/V)_{min} = \frac{y_b - y_a}{x_b^* - x_a} = \frac{0.02 - 0.003}{0.02857 - 0} = 0.595$$

Note that $x_b^* = y_b/K = 0.02/0.7 = 0.02857$. The actual L/V is $1.6(L/V)_{min}$. Consequently

$$L/V = 1.6 \times 0.595 = 0.952$$

The mole fraction of ethyl alcohol in the liquid leaving the column at the bottom, that is, x_b , can be calculated from the operating line using the terminal points:

$$y_b = y_a + (\frac{L}{V})(x_b - x_a)$$

$$0.02 = 0.003 + (0.952\ )(x_b - 0)$$

$$x_b = 0.01786$$

Thus, the actual operating line passes through the points (0, 0.003) and (0.01786, 0.02).

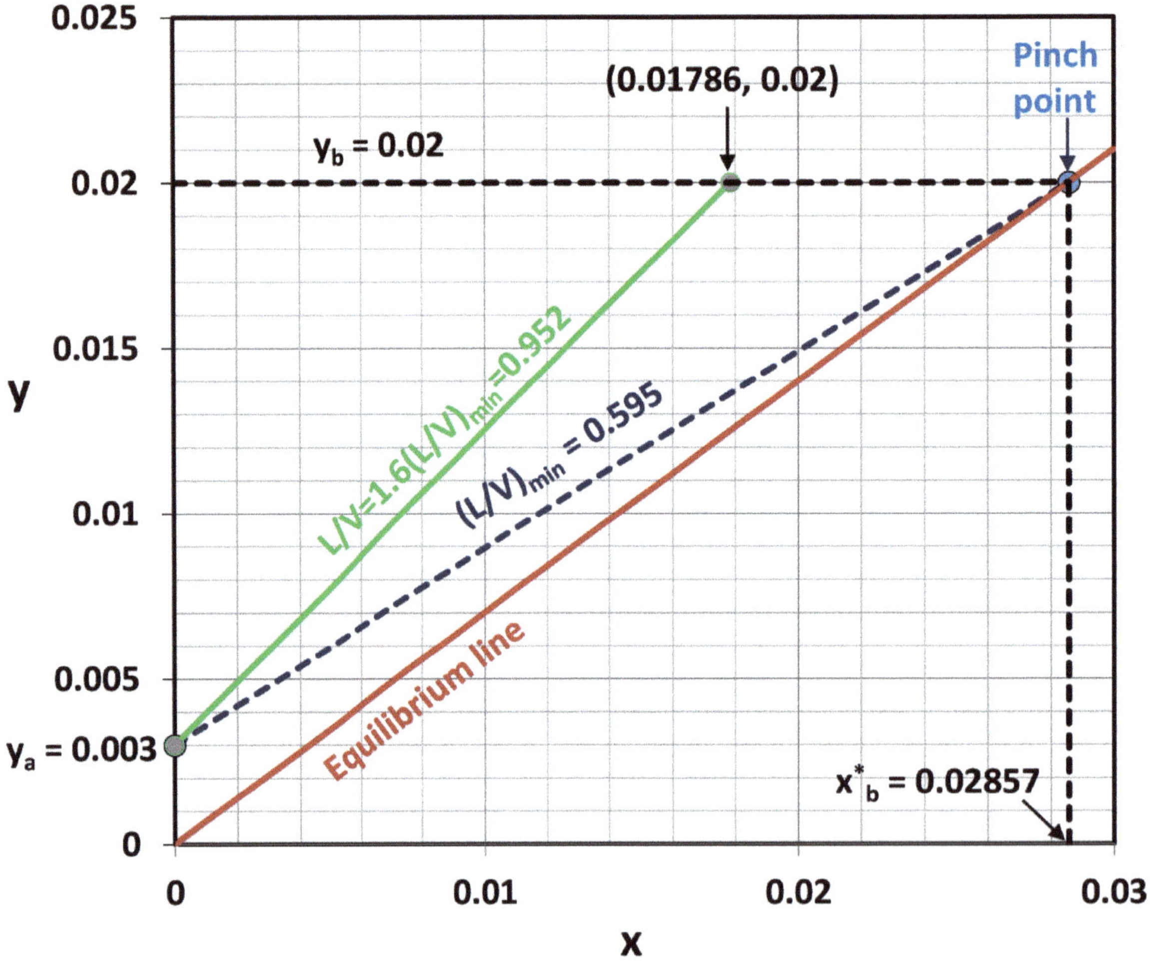

The stepping off the stages using the McCabe-Thiele method is shown in the figure below. The number of equilibrium stages required is 3.08.

b) Let us now apply the Kremser equation to estimate the number of equilibrium stages. The absorption factor is

$$A = (L/V)/K = 0.952/0.70 = 1.36$$

As $A \neq 1$, the following form of Kremser equation is applicable:

$$N = \frac{ln\left(\frac{y_b - y_b^*}{y_a - y_a^*}\right)}{ln(A)} \qquad when\ A \neq 1$$

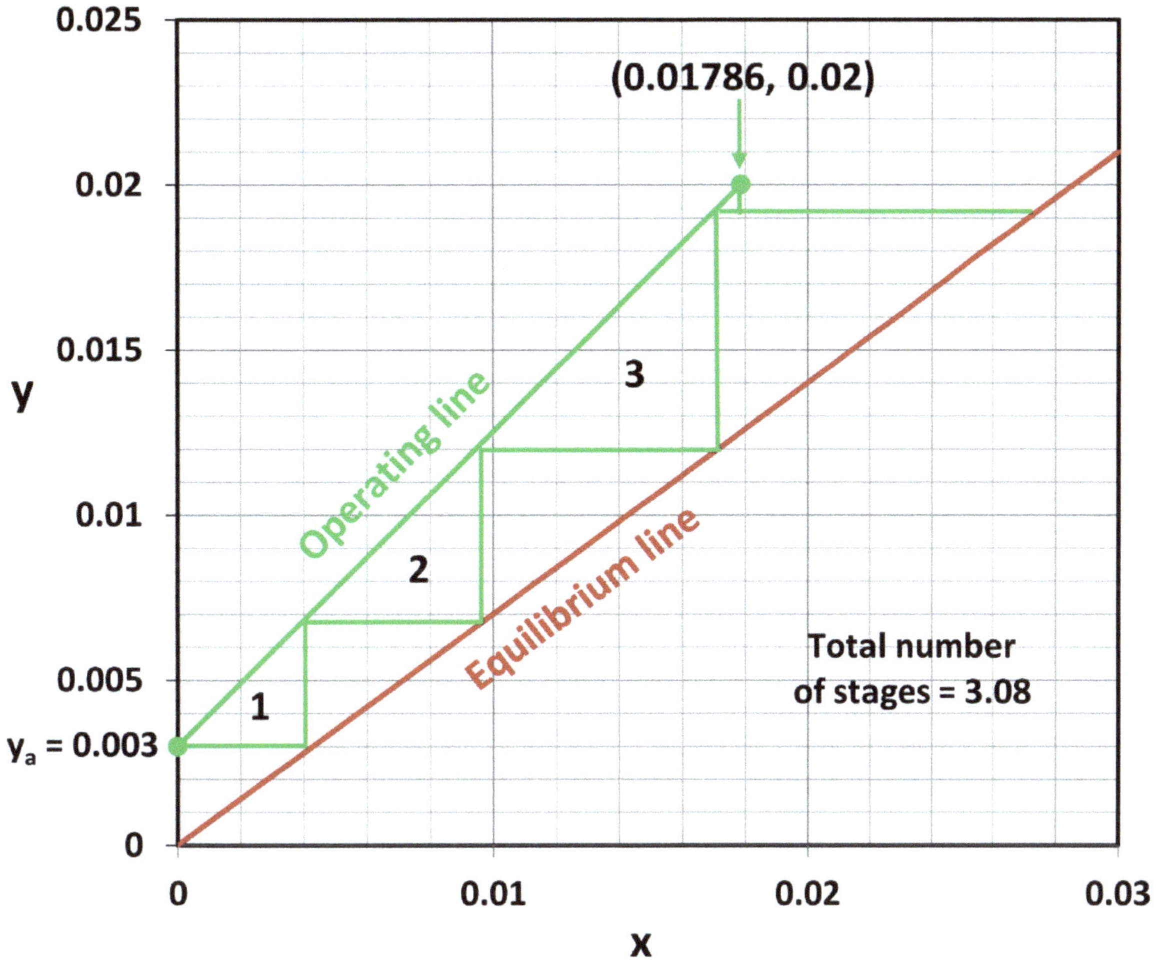

$$y_b = 0.02, \ \ y_b^* = Kx_b = 0.7 \times 0.01786 = 0.0125$$

$$y_a = 0.003, \ \ y_a^* = Kx_a = 0.7 \times 0 = 0$$

Thus,

$$N = \frac{ln\left(\frac{y_b - y_b^*}{y_a - y_a^*}\right)}{ln(A)} = \frac{ln\left(\frac{0.02-0.0125}{0.003-0}\right)}{\ln(1.36)} = 2.98$$

This value is close to the value obtained graphically using the McCabe-Thiele method.

Chapter 5

Multi-Stage Liquid Stripping

Problem 1

A hydrocarbon mixture consisting of 5 mole% propane and 95 mole% non-volatile hydrocarbon is being stripped by pure superheated steam using a multi-stage countercurrent stripping column. The ratio of steam molar flow rate to hydrocarbon molar flow rate is 1/30, that is, $V/L = 1/30$. The propane content of the discharge liquid stream is 0.1 mole%. The equilibrium relation is $y = 30x$, where y and x are mole fractions of propane in steam and hydrocarbon mixture, respectively.

a) Determine the mole fraction of propane in the exit gas stream
b) How many equilibrium stages are required using the graphical McCabe-Thiele method?
c) How many equilibrium stages are required using the Kremser equation?
You can assume steam is an inert gas and it does not condense.

Solution

a) Refer to the column schematic shown below.

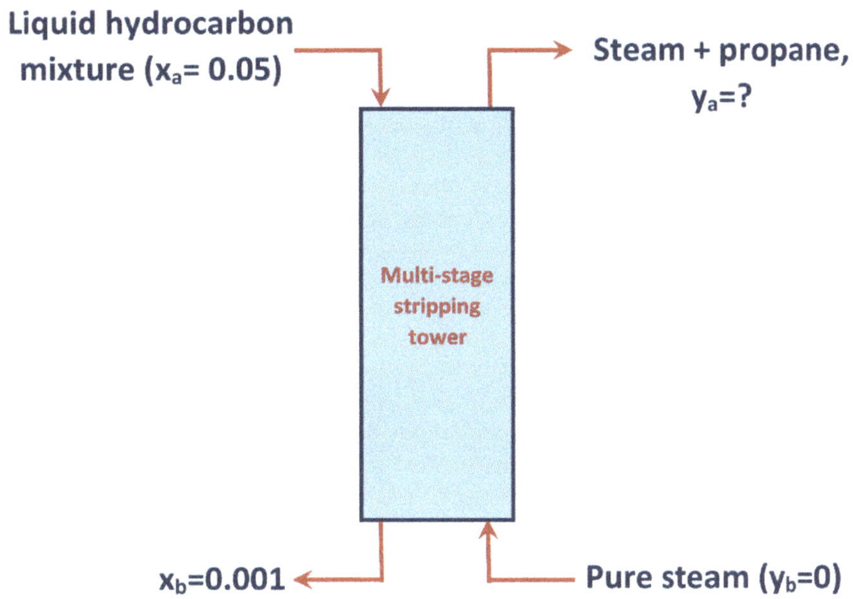

Let us first treat the system as dilute. The mole fraction of solute in the gas stream leaving the column at the top, that is, y_a , can be calculated from the operating line using the terminal points as:

$$y_b = y_a + (\frac{L}{V})(x_b - x_a)$$

$$0 = y_a + (30)(0.001 - 0.05) \;;\; y_a = 1.45$$

This is impossible as mole fraction cannot exceed unity. The problem with this solution is that L/V is assumed to be constant. This assumption is incorrect.

Let us take the molar flow rate of liquid hydrocarbon mixture at the entrance as 1 kmol/h as the basis of calculations. As $V/L = 1/30$, $V = 1/30$ kmol/h at the entrance of the column.

$$L' = L_a(1 - x_a) = 1(1 - 0.05) = 0.95 \text{ kmol/h}$$

$$L_b = L'/(1 - x_b) = 0.95/(1 - 0.001) = 0.95095 \text{ kmol/h}$$

The total liquid molar flow rate has decreased from 1 kmol/h to 0.95095 kmol/h. The amount of propane transferred from liquid phase to gas phase is (1-0.95095)=0.04905 kmol/h. Consequently,

$$V_a = V_b + 0.04905 = (1/30) + 0.04905 = 0.082383 \text{ kmol/h}$$

Thus the gas phase molar flow rate has increased from 0.03333 kmol/h at the entrance to 0.082383 kmol/h at the exit (nearly 2.5 times). In conclusion, neither the liquid molar flow rate nor the gas molar flow rate is constant in the column. Hence we must treat the system as concentrated and use the mole ratios in our analysis. The solution is as follows:

$$L' = 0.95 \text{ kmol/h}$$

$$V' = V_b = 1/30 = 0.0333 \text{ kmol/h}$$

$$X_a = \frac{x_a}{1 - x_a} = \frac{0.05}{1 - 0.05} = 0.05263$$

$$X_b = \frac{x_b}{1 - x_b} = \frac{0.001}{1 - 0.001} = 0.001$$

$$Y_b = \frac{y_b}{1 - y_b} = 0$$

The operating line on Y-X scale is:

$$Y = Y_a + \left(\frac{L'}{V'}\right)(X - X_a)$$

When applied at the bottom of the column, the operating line gives:

$$Y_b = Y_a + \left(\frac{L'}{V'}\right)(X_b - X_a)$$

$$0 = Y_a + \left(\frac{0.95}{0.0333}\right)(0.001 - 0.05263)$$

$$Y_a = 1.47145$$

$$y_a = \frac{Y_a}{1 + Y_a} = \frac{1.47145}{1 + 1.47145} = 0.59538$$

ANS: The mole fraction of propane in the exit gas stream is 0.5954.

b) The number of equilibrium stages required using the graphical McCabe-Thiele method?

We need to draw the operating line and equilibrium curve on Y-X graph.

The operating line on Y-X scale is a straight line passing through the terminal points (0.001, 0) and (0.0526, 1.471).

The equilibrium curve on Y-X scale follows from the given equilibrium relation on y-x scale:

$$y = 30x \;;\quad \frac{Y}{1+Y} = 30\left(\frac{X}{1+X}\right)$$

The following figure shows the plots of the operating and equilibrium lines. The stepping off the stages using the McCabe-Thiele method is also shown in the figure. The number of equilibrium stages required is 7.0.

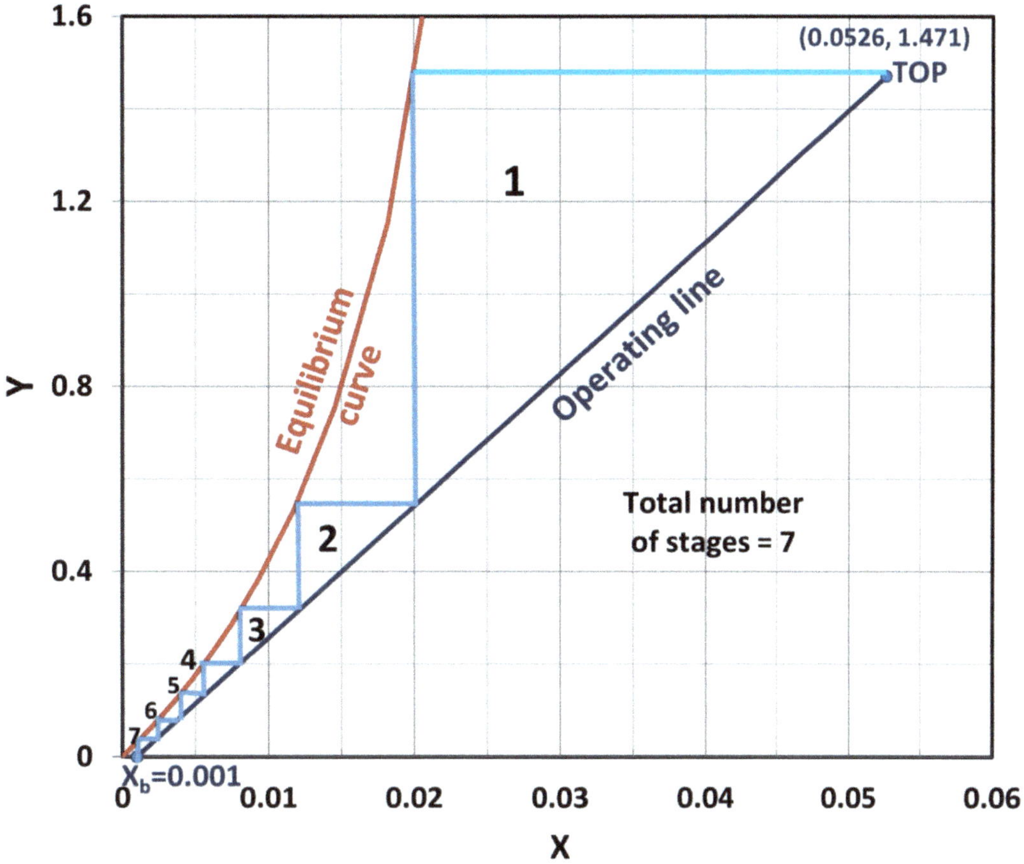

c) The number of equilibrium stages using the Kremser equation?

The Kremser equation cannot be applied in the present case as the equilibrium relation on Y-X scale, given below, is not linear:

$$\frac{Y}{1+Y} = 30\left(\frac{X}{1+X}\right)$$

Problem 2

A multi-stage countercurrent stripping tower is being used to remove ammonia from wastewater using pure air as the stripping gas. The water entering the process contains 0.05 mole fraction ammonia. The ammonia mole fraction in the exit water stream is 0.01. The equilibrium relation is $y = 1.5x$, where y and x are mole fractions of ammonia in air and water streams at equilibrium, respectively. The molar flow rate of air is 1.5 times the molar flow rate of wastewater.

a) Determine the mole fraction of ammonia in the exit gas stream
b) How many equilibrium stages are required using the graphical McCabe-Thiele method?
c) How many equilibrium stages are required using the Kremser equation?

You can assume that the molar flow rates of gas and liquid streams are constant.

Solution

a) Refer to the column schematic shown below.

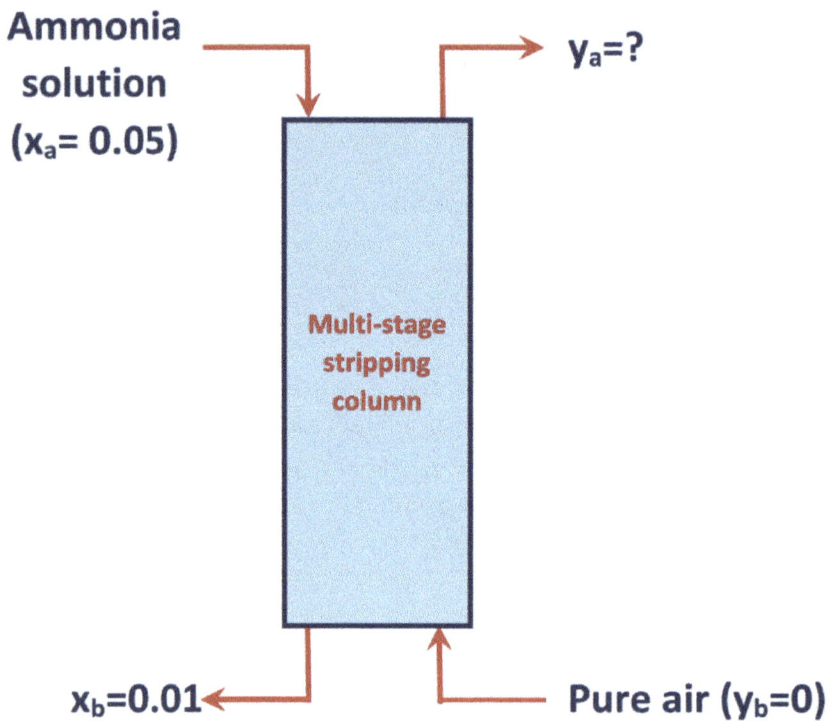

The mole fraction of solute in the gas stream leaving the column at the top, that is, y_a , can be calculated from the operating line using the terminal points:

$$y_b = y_a + (\frac{L}{V})(x_b - x_a)$$

$$0 = y_a + (1/1.5\,)(0.01 - 0.05)$$

$$y_a = 0.0267$$

b) Equilibrium stages required using the graphical McCabe-Thiele method?

The operating line passes through the terminal points (0.05, 0.0267) and (0.01, 0). The following figure shows the plots of the operating and equilibrium lines. The stepping off the stages using the McCabe-Thiele method is also shown in the figure. The number of equilibrium stages required is 1.6.

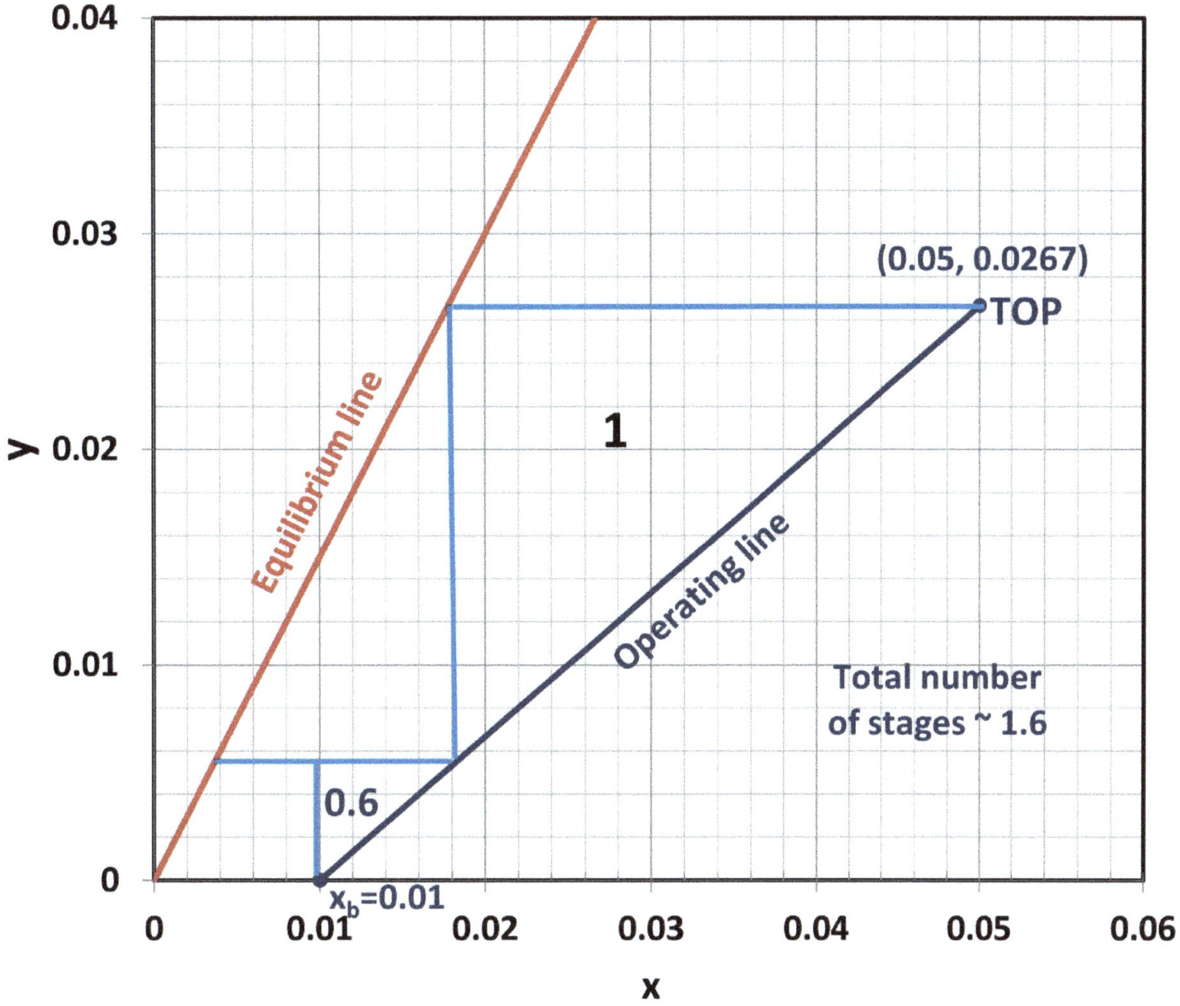

c) Let us now apply the Kremser equation to estimate the number of equilibrium stages. The stripping factor is

$$S = K/(L/V) = 1.5 \times 1.5 = 2.25$$

As $S \neq 1$, the following form of Kremser equation is applicable:

$$N = \frac{\ln\left(\frac{x_a - x_a^*}{x_b - x_b^*}\right)}{\ln(S)} = \frac{\ln\left(\frac{0.05-0.0178}{0.01-0}\right)}{\ln(2.25)} = 1.44$$

Note that:

$$x_a^* = y_a/K = 0.0267/1.5 = 0.0178$$

$$x_b^* = y_b/K = 0/1.5 = 0$$

The number of equilibrium stages estimated from the Kremser equation is close to the value obtained graphically from the McCabe-Thiele method.

Chapter 6

Some Definitions and Basic Concepts related to Distillation

Problem 1

The normal boiling point of a certain liquid is 35°C. Calculate the vapor pressure at 30°C if the material follows the Trouton rule. Note that according to the Trouton rule, $\Delta H_n^{vap} = 10.5RT_n$ where T_n is the normal boiling point, ΔH_n^{vap} is the latent heat of vaporization at normal boiling point, and R is the universal gas constant. You can assume that the latent heat of vaporization is constant independent of temperature.

Solution

$$T_n = 35^oC = 308.15\ K; \qquad P^{Sat} = 101.325\ kPa$$

$$\Delta H_n^{vap} = 10.5RT_n = 10.5 \times 8.314 \times 308.15 = 26{,}900.6\ J/mole$$

The Clausius-Clapeyron equation can be written as:

$$ln\left(\frac{P_2^{Sat}}{P_1^{Sat}}\right) = -\frac{\Delta H^{vap}}{R}\left(\frac{1}{T_2} - \frac{1}{T_1}\right)$$

We need to calculate P_2^{Sat} given that: $T_1 = 35^oC = 308.15\ K$, $P_1^{Sat} = 101.325\ kPa$, $T_2 = 30^oC = 303.15\ K$. Thus:

$$ln\left(\frac{P_2^{Sat}}{101.325}\right) = -\frac{26900.6}{8.314}\left(\frac{1}{303.15} - \frac{1}{308.15}\right)$$

The equation gives $P_2^{Sat} = 85.213\ kPa$.

Problem 2

100 g of n-hexane (MW=86.18 g/mole) is mixed with 50 g of n-octane (MW=114.23 g/mole) to form a liquid solution. The normal boiling point of n-hexane is 68.7°C and that of n-octane is 125.7°C. The latent heats of vaporization of n-hexane and n-octane are: 335 kJ/kg and 303 kJ/kg, respectively.

a) What are the partial pressures of n-hexane and n-octane above the solution at 25°C? What is the total pressure?
b) What is the composition of the vapor mixture above the liquid solution at 25°C?

Assume liquid solution and vapor mixture are ideal in nature.

Solution

The liquid solution consists of 100 g (=100/86.18=1.1604 moles) of n-hexane and 50 g (=50/114.23=0.4377 moles) of n-octane. Thus the mole fractions of n-hexane and n-octane in the liquid solution are as follows:

$$x_{hexane} = 0.7261 \quad \text{and} \quad x_{octane} = 0.2739$$

The latent heats of vaporization of n-hexane and n-octane are: 335 kJ/kg (=335x86.18=28,870.3 J/mole) and 303 kJ/kg (=303x114.23=34,611.69 J/mole), respectively.

The vapor pressures of pure n-hexane and n-octane can be estimated from the Clausius-Clapeyron equation:

$$ln\left(\frac{P_2^{Sat}}{P_1^{Sat}}\right) = -\frac{\Delta H^{vap}}{R}\left(\frac{1}{T_2} - \frac{1}{T_1}\right)$$

For n-hexane, $T_1 = 68.7^oC = 341.85\ K,\ P_1^{Sat} = 1\ atm,\ T_2 = 25^oC = 298.15\ K,\ P_2^{Sat} =?$ Thus:

$$ln\left(\frac{P_2^{Sat}}{1}\right) = -\frac{28870.3}{8.314}\left(\frac{1}{298.15} - \frac{1}{341.85}\right)$$

The equation gives $P_2^{Sat} = 0.22563\ atm.$

For n-octane, $T_1 = 125.7^oC = 398.85\ K,\ P_1^{Sat} = 1\ atm,\ T_2 = 25^oC = 298.15\ K,\ P_2^{Sat} =?$ Thus:

$$ln\left(\frac{P_2^{Sat}}{1}\right) = -\frac{34611.7}{8.314}\left(\frac{1}{298.15} - \frac{1}{398.85}\right)$$

The equation gives $P_2^{Sat} = 0.02944\ atm.$

a) Partial pressures and total pressure:

$$P_{hexane} = x_{hexane}P_{hexane}^{Sat} = 0.7261 \times 0.22563 = 0.16383 \text{ atm}$$

$$P_{octane} = x_{octane}P_{octane}^{Sat} = 0.2739 \times 0.02944 = 0.08064 \text{ atm}$$

$$total\ pressure = P_{hexane} + P_{octane} = 0.16383 + 0.08064 = 0.24447 \text{ atm}$$

b) Composition of vapor mixture:

$$y_{hexane}P = P_{hexane} = 0.16383$$

$$y_{hexane} = 0.16383/0.24447 = 0.6701$$

$$y_{octane}P = P_{octane} = 0.08064$$

$$y_{octane} = 0.08064/0.24447 = 0.3299$$

$$\sum y_i = 0.6701 + 0.3299 = 1.0$$

Problem 3

Estimate the relative volatility of n-hexane with respect to n-octane in a binary n-hexane/n-octane system at different temperatures: a) 80 °C, b) 90 °C, c) 100 °C, and d) 120 °C. Assume that the system is ideal in nature and Raoult's law is applicable. The Antoine equations for n-hexane and n-octane are as follows:

$$lnP_{hexane}^{Sat} = 13.8193 - \frac{2696.04}{t+224.317}$$

$$lnP_{octane}^{Sat} = 13.9346 - \frac{3123.13}{t+209.635}$$

where t is the temperature in °C and P^{Sat} is vapor pressure in kPa.

Solution

For ideal systems,

$$\alpha_{ij} = \frac{P_i^{Sat}}{P_j^{Sat}}$$

Thus,

$$\alpha_{hexane-octane} = \frac{P_{hexane}^{Sat}}{P_{octane}^{Sat}}$$

Using Antoine equations, we get:

T(°C)	P_{hexane}^{Sat}(kPa)	P_{octane}^{Sat}(kPa)	$\alpha_{hexane-octane}$
80	142.59	23.37	6.1
90	189.02	33.50	5.64
100	246.24	46.90	5.25
120	399.09	86.50	4.61

Comments: The relative volatility decreases somewhat with the increase in temperature. However, the decrease is less than 25% over the temperature increase of 50% from 80 to 120 °C.

Problem 4

Consider a 4-component liquid solution containing 25 mole% species 1, 25 mole% species 2, 25 mole% species 3, and 25 mole% species 4 at a pressure of 1 bar. For this system, the vapor pressures of the pure species are given by the Antoine equation:

$$lnP_i^{Sat} = A_i - \frac{B_i}{T+C_i}$$

where T is in K and the vapor pressures are in bars. The Antoine coefficients are given in the following table:

Species	1	2	3	4
A_i	9.2131	9.1325	9.2164	9.2535
B_i	2477.07	2766.63	2697.55	2911.32
C_i	-39.94	-50.50	-48.78	-56.51

a) Determine the bubble point temperature of this liquid solution.
b) What is the composition of the first vapor bubble that is formed?
You can assume that the liquid solution and vapor mixture are ideal in nature. In your calculations, use species 3 as the reference species.

Solution

This problem involves BUBLT calculation. We are given liquid composition and P, we are asked to determine T and bubble composition.

a) Estimation of bubble point temperature:

$$T_{guess} = \sum x_i T_i^{Sat}$$

At P = 1 bar, Antoine equations give: $T_1^{Sat} = 308.9039\,K$, $T_2^{Sat} = 353.4433\,K$, $T_3^{Sat} = 341.4702\,K$, $T_4^{Sat} = 371.1283\,K$. Thus, $T_{guess} = 343.7114\,K$.

At $T = 343.7114\,K,\ \alpha_{13} = 2.6878, \alpha_{23} = 0.6886,\ \alpha_{33} = 1.0, \alpha_{43} = 0.3854$. Using Equation (6.25):

$$P_3^{Sat} = P/\sum x_i\,\alpha_{ir}$$

$$= 1/(0.25 \times 2.6878 + 0.25 \times 0.6886 + 0.25 \times 1 + 0.25 \times 0.3854) = 0.84\ bar$$

Using Equation (6.27), we obtain:

$$T = \frac{B_3}{A_3 - lnP_3^{Sat}} - C_3 = 336.04\ K$$

Since T of 336.04 K is significantly off from the assumed value of 343.7114 K, we need another iteration.

At $T = 336.04\,K,\ \alpha_{13} = 2.78, \alpha_{23} = 0.682,\ \alpha_{33} = 1.0, \alpha_{43} = 0.373$. Using Equation (6.25):

$$P_3^{Sat} = P/\sum x_i\,\alpha_{ir}$$

$$= 1/(0.25 \times 2.78 + 0.25 \times 0.682 + 0.25 \times 1 + 0.25 \times 0.373) = 0.83\ bar$$

Using Equation (6.27), we obtain:

$$T = \frac{B_3}{A_3 - lnP_3^{Sat}} - C_3 = 335.6\ K$$

This value of T is very close to the assumed value of 336.04 K. Thus T=335.6 K

b) Composition of vapor bubble: The composition of the first vapor bubble is determined from Raoult's law as:

$$y_1 = x_1 P_1^{Sat}/P = 0.25 \times 2.305/1 = 0.5762$$

$$y_2 = x_2 P_2^{Sat}/P = 0.25 \times 0.5646/1 = 0.1412$$

$$y_3 = x_3 P_3^{Sat}/P = 0.25 \times 0.8281/1 = 0.2070$$

$$y_4 = x_4 P_4^{Sat}/P = 0.25 \times 0.3079/1 = 0.077$$

Note that at T=335.6 K, the vapor pressures of pure species are determined from the Antoine equations.

Problem 5

Consider a 3-component system of species 1, 2 and 3. For this system, the vapor pressures of the pure species are given by the Antoine equations as:

$$lnP_1^{Sat} = 13.7819 - \frac{2726.81}{t + 217.572}$$

$$lnP_2^{Sat} = 13.932 - \frac{3056.96}{t+217.625}$$

$$lnP_3^{Sat} = 13.9726 - \frac{3259.93}{t+212.30}$$

where t is in °C and the vapor pressures are in kPa. Species 1 is more volatile.

Assuming that the system is ideal in nature and follows the Raoult law, do the following calculations:
a) BUBLP calculation. Given that $t = 120^oC,\ x_1 = 0.25, x_2 = 0.35, x_3 = 0.40$
b) DEWP calculation. Given that $t = 110^oC,\ y_1 = 0.40, y_2 = 0.35, y_3 = 0.25$
c) BUBLT calculation. Given that $P = 95\ kPa,\ x_1 = 0.45, x_2 = 0.20, x_3 = 0.35$
d) DEWT calculation. Given that $P = 100\ kPa,\ y_1 = 0.50, y_2 = 0.30, y_3 = 0.20$

Solution

a) BUBLP calculation. Given that $t = 120^oC,\ x_1 = 0.25, x_2 = 0.35, x_3 = 0.40$

$$P = \sum x_i\, P_i^{Sat}$$

Species	A_i	B_i	C_i	T(°C)	P_i^{Sat}(kPa)
1	13.7819	2726.81	217.572	120	300.1211
2	13.932	3056.96	217.625	120	131.3275
3	13.9726	3259.93	212.3	120	64.2254

$P = x_1P_1^{Sat} + x_2P_2^{Sat} + x_3P_3^{Sat} = 0.25 \times 300.1211 + 0.35 \times 131.3275 + 0.40 \times 64.2254 =$ 146.6851 kPa

Composition of first vapor bubble:

$$y_1 = x_1P_1^{Sat}/P = 0.25 \times 300.1211/146.6851 = 0.5115$$

$$y_2 = x_2P_2^{Sat}/P = 0.35 \times 131.3275/146.6851 = 0.3134$$

$$y_3 = x_3P_3^{Sat}/P = 0.45 \times 64.2254/146.6851 = 0.1751$$

b) DEWP calculation. Given that $t = 110^oC,\ y_1 = 0.40, y_2 = 0.35, y_3 = 0.25$

$$P = 1/\sum y_i/P_i^{Sat}$$

Species	A_i	B_i	C_i	T(°C)	P_i^{Sat}(kPa)
1	13.7819	2726.81	217.572	110	234.5321
2	13.932	3056.96	217.625	110	99.617
3	13.9726	3259.93	212.3	110	47.3713

$$P = 1/\left(y_1/P_1^{Sat} + y_2/P_2^{Sat} + y_3/P_3^{Sat}\right) = 1/[(0.40/234.5321) + (0.35/99.617) + (0.25/47.3713)] = 95.2705 \text{ kPa}$$

Composition of the first liquid drop:

$$x_1 = y_1P/P_1^{Sat} = 0.40 \times 95.2705/234.5321 = 0.1625$$

$$x_2 = y_2P/P_2^{Sat} = 0.35 \times 95.2705/99.617 = 0.3347$$

$$x_3 = y_3P/P_3^{Sat} = 0.25 \times 95.2705/47.3713 = 0.5028$$

c) BUBLT calculation. Given that $P = 95\ kPa,\ x_1 = 0.45, x_2 = 0.20, x_3 = 0.35$

We are given liquid composition and P, we are required to determine the bubble point T and initial bubble composition. We follow the trial-and-error procedure discussed in section 6.6.3

Initial guess of bubble point temperature:

$$T_{guess} = \sum x_i\, T_i^{Sat}$$

At P = 95 kPa, Antoine equations give: $T_1^{Sat} = 77.9203^oC,\ T_2^{Sat} = 108.3421^oC,\ T_3^{Sat} = 133.8117^oC.$ Thus, $T_{guess} = 103.5667^oC.$

At $T = 103.5667^oC,\ \alpha_{11} = 1.0, \alpha_{21} = 0.4163,\ \alpha_{31} = 0.1942.$ Using Equation (6.25):

$$P_1^{Sat} = P/\sum x_i\, \alpha_{ir}$$

$$= 95/(0.45 \times 1 + 0.20 \times 0.4163 + 0.35 \times 0.1942) = 158.0095 \text{ kPa}$$

Using Equation (6.27), we obtain:

$$T = \frac{B_1}{A_1 - lnP_1^{Sat}} - C_1 = 95.163^oC$$

Since T of 95.163^oC is significantly off from the assumed value of 103.5667^oC, we need another iteration.

At $T = 95.163^oC,\ \alpha_{11} = 1.0, \alpha_{21} = 0.405,\ \alpha_{31} = 0.1840.$ Using Equation (6.25):

$$P_1^{Sat} = P/\sum x_i\, \alpha_{ir}$$

$$= 95/(0.45 \times 1 + 0.20 \times 0.405 + 0.35 \times 0.1840) = 159.56 \text{ kPa}$$

$$T = \frac{B_1}{A_1 - lnP_1^{Sat}} - C_1 = 95.513^oC$$

This value of T is very close to the assumed value of 95.163^oC. Thus T=95.513^oC.

Composition of vapor bubble:

$$y_1 = x_1P_1^{Sat}/P = 0.45 \times 159.5454/95 = 0.7557$$

$$y_2 = x_2P_2^{Sat}/P = 0.20 \times 64.6873/95 = 0.1362$$

$$y_3 = x_3P_3^{Sat}/P = 0.35 \times 29.426/95 = 0.1084$$

Note that at T=95.513^oC, the vapor pressures of pure species are determined from the Antoine equations.

d) DEWT calculation. Given that $P = 100\ kPa,\ y_1 = 0.50, y_2 = 0.30, y_3 = 0.20$

We are given vapor composition and P, we are required to determine T and dew (first liquid drop) composition. We follow the trial-and-error procedure discussed in section 6.6.3.
Initial guess of dew point temperature:

$$T_{guess} = \sum y_i T_i^{Sat}$$

At P = 100 kPa, Antoine equations give: $T_1^{Sat} = 79.572^oC,\ T_2^{Sat} = 110.135^oC,\ T_3^{Sat} = 135.707^oC$. Thus, $T_{guess} = 99.9678^oC$.

At $T = 99.9678^oC,\ \alpha_{11} = 1.0, \alpha_{21} = 0.4115,\ \alpha_{31} = 0.1898$. Using Equation (6.31):

$$P_1^{Sat} = P\sum y_i/\alpha_{i3} = 1 \times (y_1/\alpha_{13} + y_2/\alpha_{23} + y_3/\alpha_{33} + y_4/\alpha_{43}) = 228.26 \text{ kPa}$$

Using Equation (6.27), we obtain:

$$T = \frac{B_1}{A_1 - lnP_1^{Sat}} - C_1 = 108.936^oC$$

Since T of 108.936^oC is significantly off from the assumed value of 99.9678^oC, we need another iteration.

At $T = 108.936^oC,\ \alpha_{11} = 1.0, \alpha_{21} = 0.4234,\ \alpha_{31} = 0.2007$. Using Equation (6.31):

$$P_1^{Sat} = P\sum y_i/\alpha_{i3} = 1 \times (y_1/\alpha_{13} + y_2/\alpha_{23} + y_3/\alpha_{33} + y_4/\alpha_{43}) = 220.514 \text{ kPa}$$

Using Equation (6.27), we obtain:

$$T = \frac{B_1}{A_1 - lnP_1^{Sat}} - C_1 = 107.5925^oC$$

This value of T is very close to the assumed value of 108.936^oC. Thus T=107.5925^oC.

Composition of liquid droplet:

$$x_1 = y_1P/P_1^{Sat} = 0.50 \times 100/220.514 = 0.2267$$

$$x_2 = y_2P/P_2^{Sat} = 0.30 \times 100/92.9684 = 0.3227$$

$$x_3 = y_3P/P_3^{Sat} = 0.20 \times 100/43.8991 = 0.4556$$

Note that at T=107.5925^oC, the vapor pressures of pure species are determined from the Antoine equations.

Problem 6

Consider a 4-component mixture of n-butane (35 mole%), n-pentane (30 mole %), n-hexane (15 mole%) and n-heptane (20 mole%) at a pressure of 500 kPa. The K_i values of different species can be estimated from Equation (6.12) using Table 6.1 data at any given temperature and pressure.

Assuming that the mixture is in liquid state, estimate the bubble point temperature of the liquid? What is the composition of the first vapor bubble formed?

Solution

BUBLT calculation. Given that $P = 500\ kPa,\ x_1 = 0.35,\ x_2 = 0.30,\ x_3 = 0.15,\ x_4 = 0.20.$

Initial guess of the bubble point temperature can be made in two different ways: Antoine equation method and K- method.

Antoine equation method:

$$T_{guess} = \sum x_i\, T_i^{Sat}$$

where T_i^{Sat} is the saturation temperature of species i obtained from the Antoine equation at given total pressure of 500 kPa. The Antoine equation is:

$$lnP_i^{Sat} = A_i - \frac{B_i}{T+C_i}$$

where T is in °C and the vapor pressures are in kPa. The Antoine coefficients are given in the following table:

Species	n-butane (species 1)	n-pentane (species 2)	n-hexane (species 3)	n-heptane (species 4)
A_i	9.2131	9.1325	9.2164	9.2535
B_i	2477.07	2766.63	2697.55	2911.32
C_i	-39.94	-50.50	-48.78	-56.51

At P = 500 kPa, Antoine equations give: $T_1^{Sat} = 50.5804^oC,\ T_2^{Sat} = 92.6484^oC,\ T_3^{Sat} = 130.2062^oC,$ $T_4^{Sat} = 164.114^oC$. Thus, $T_{guess} = 97.85^oC$.

K - method:

$$T_{guess} = \sum x_i\, T_i^{Sat}$$

where T_i^{Sat} is the saturation temperature of species i obtained from the DePriester charts or equivalently Equation (6.12) taking $K_i = 1.0$. Equation (6.12) reduces to the following when $K_i = 1.0$:

$$T_i^{Sat} = \sqrt{\frac{a_1}{-a_o - b_o lnP}}$$

where a_o , a_1, and b_o are given as follows for the species under consideration (also see Table 6.1 of the book):

Species	n-butane (species 1)	n-pentane (species 2)	n-hexane (species 3)	n-heptane (species 4)
a_o	7.994986	7.33129	6.96783	6.52914
a_1	-1280557	-1524891	-1778901	-2013803
b_o	-0.96455	-0.89143	-0.84634	-0.79543

At P=72.5189 psia (=500 kPa), the above equation gives: $T_1^{Sat} = 575.7541^oR$, $T_2^{Sat} = 658.8837^oR$, $T_3^{Sat} = 729.5535^oR$, $T_4^{Sat} = 803.1878^oR$. Thus, $T_{guess} = 669.2496^oR = 98.47^oC$.

Both methods give similar values of T_{guess}. We take $T_{guess} = 97.85^oC = 668.13^oR$ in the following calculations.

At $T = 668.13^oR$ and P=72.5189 psia, Equation (6.12) gives the following values of K_i: $K_1 = 2.703$, $K_2 = 1.1014$, $K_3 = 0.5258$, $K_4 = 0.2492$. Thus the values of α_{ir} (species 1 being the reference species) are: $\alpha_{11} = 1.0, \alpha_{21} = 0.4075$, $\alpha_{31} = 0.1945$, $\alpha_{41} = 0.0184$. Using Equation (6.36),

$$K_r = 1/\sum x_i\,\alpha_{ir}$$

$$= 1/(0.35 \times 1 + 0.30 \times 0.4075 + 0.15 \times 0.1945 + 0.20 \times 0.0922) = 1.9236$$

Using Equation (6.12), the improved value of temperature is obtained as $T = 631.725^oR$. Since T of 631.725^oR is significantly off from the assumed value of 668.13^oR, another iteration is needed.

At $T = 631.725^oR$, and P=72.5189 psia, Equation (6.12) gives the following values of K_i: $K_1 = 1.9237$, $K_2 = 0.7346$, $K_3 = 0.3278$, $K_4 = 0.146$. Thus the values of α_{ir} (species 1 being the reference species) are: $\alpha_{11} = 1.0, \alpha_{21} = 0.3819$, $\alpha_{31} = 0.1704$, $\alpha_{41} = 0.0759$. Using Equation (6.36),

$$K_r = 1/\sum x_i\,\alpha_{ir}$$

$$= 1/(0.35 \times 1 + 0.30 \times 0.3819 + 0.15 \times 0.1704 + 0.20 \times 0.0759) = 1.979$$

Using Equation (6.12), the improved value of temperature is obtained as $T = 634.54^oR$. This value of T is very close to the assumed value of 631.725^oR . Thus the bubble point temperature is T=$634.54^oR = 79.2^oC$.

Composition of vapor bubble:

$$y_1 = K_1x_1 = 1.979 \times 0.35 = 0.6926$$

$$y_2 = K_2x_2 = 0.7598 \times 0.30 = 0.2279$$

$$y_3 = K_3x_3 = 0.341 \times 0.15 = 0.0511$$

$$y_4 = K_4x_4 = 0.1526 \times 0.20 = 0.0305$$

Note that at T=634.54^oR and P=72.5189 psia, the K_i's of various species are determined from Equation (6.12).

Problem 7

Consider a binary system of species 1 and 2 at a total pressure of 101.32 kPa. Assuming that the system is ideal in nature and follows the Raoult law, construct the T-x-y and y-x diagrams for the system at 101.32 kPa. For this system, the vapor pressures of the pure species are given by the Antoine equations as:

$$lnP_1^{Sat} = 13.8193 - \frac{2696.04}{t+224.317}$$

$$lnP_2^{Sat} = 13.9346 - \frac{3123.13}{t+209.635}$$

where t is in °C and the vapor pressures are in kPa. Species 1 is more volatile.
When a liquid solution containing 25 mole% species 1 is heated at a constant pressure of 101.32 kPa, at what temperature would the first vapor bubble form and what is its composition?

Solution

The boiling points of pure species 1 and 2 at 101.32 kPa are calculated from the given Antoine equations as: $T_1^{Sat} = 68.70^oC$ and $T_2^{Sat} = 125.60^oC$. Table given below shows at different temperatures the values of vapor pressures of pure species and mole fractions of more volatile component (species 1) in the vapor and liquid phases at equilibrium with each other. The Antoine equations are used to calculate the vapor pressures. Equations (6.44) and (6.45) are used to calculate the mole fractions in liquid and vapor phases at equilibrium.

The T-x-y and y-x diagrams for the system at 101.32 kPa are developed using the data shown in the table. Note that x and y refer to mole fractions of more volatile species (species 1). The diagrams are shown in the figures given below.

When a liquid solution containing 25 mole% species 1 is heated at a constant pressure of 101.32 kPa, the first vapor bubble is formed at 101.6 °C and the composition of the first bubble is $y = 0.635$ (Please see the T-x-y diagram).

T (°C)	$P_1^{Sat}(kPa)$	$P_2^{Sat}(kPa)$	x	y
68.7	101.32	15.09	1.00	1.00
72.7	114.69	17.69	0.86	0.98
76.7	129.39	20.64	0.74	0.95
80.7	145.51	23.99	0.64	0.91
84.7	163.15	27.76	0.54	0.87
88.7	182.40	32.01	0.46	0.83
92.7	203.34	36.76	0.39	0.78
96.7	226.07	42.07	0.32	0.72
100.7	250.68	47.98	0.26	0.65
104.7	277.28	54.54	0.21	0.57
108.7	305.96	61.79	0.16	0.49
112.7	336.82	69.79	0.12	0.39
116.7	369.96	78.59	0.08	0.28
120.7	405.47	88.25	0.04	0.16
125.60	452.34	101.32	0.00	0.00

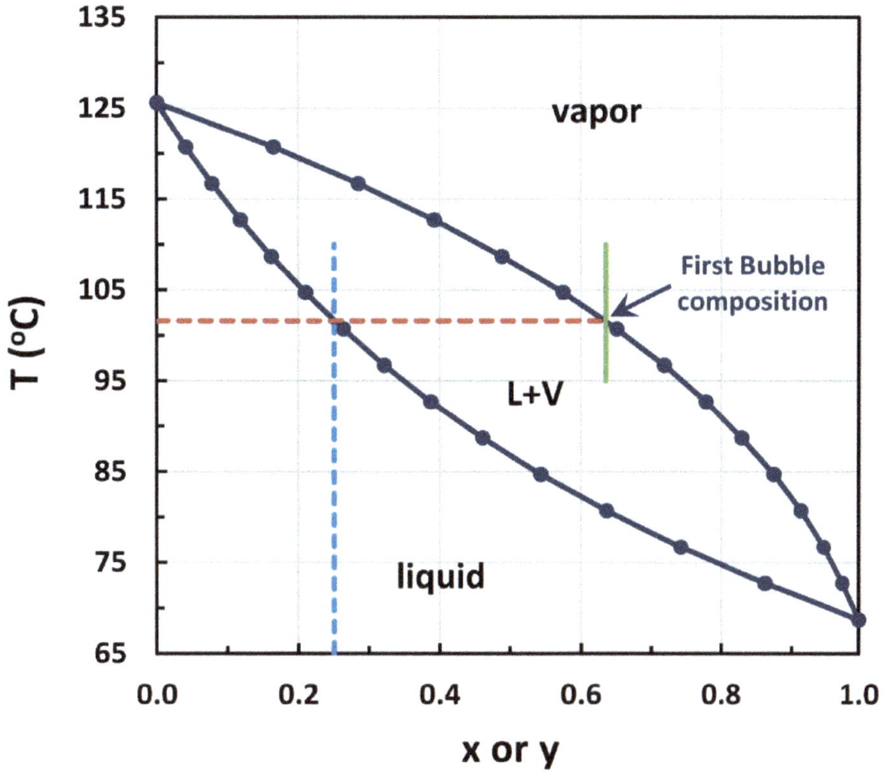
135
125
115
105
95
85
75
65
T (°C)
vapor
First Bubble
composition
L+V
liquid
0.0
0.2
0.4
0.6
0.8
1.0
x or y

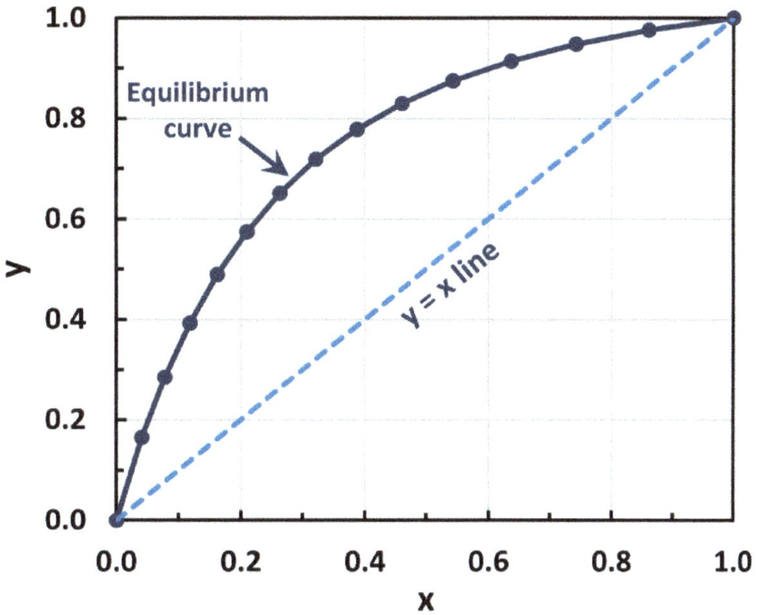
1.0
0.8
0.6
0.4
0.2
0.0
y
Equilibrium
curve
y = x line
0.0
0.2
0.4
0.6
0.8
1.0
x

Chapter 7

Mass and Energy Balances for Open (Flow) Systems

Problem 1

Water stream at 40 °C (h_1=170 kJ/kg) is mixed adiabatically with superheated steam stream at 325 °C (h_2=3063 kJ/kg) to form a saturated vapor stream (h_3=2801 kJ/kg) flowing at a mass flow rate of 9 kg/s. What are the mass flow rates of water and superheated steam entering the mixing device?

Solution

The process is shown in the figure below.

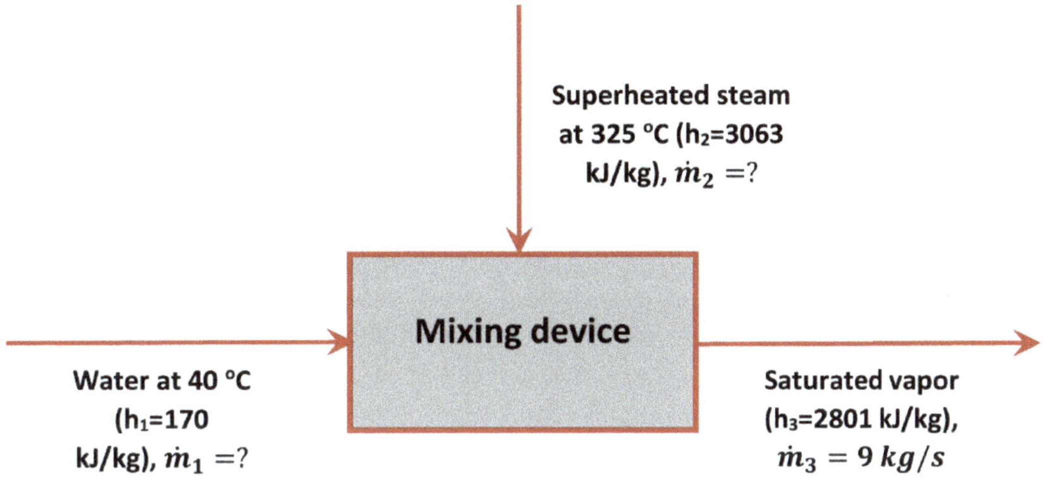

Energy balance, Equation (7.25), gives:

$$\sum_{outlets} \dot{m}_i\, h_i - \sum_{inlets} \dot{m}_i\, h_i = 0$$

$$\dot{m}_3 h_3 - (\dot{m}_1 h_1 + \dot{m}_2 h_2) = 0$$

From mass balance, Equation (7.7), we get:

$$\sum_{outlets} \dot{m}_i = \sum_{inlets} \dot{m}_i$$

$$\dot{m}_3 = \dot{m}_1 + \dot{m}_2 = 9$$

The energy balance can now be written as:

$$9(2801) - \dot{m}_1(170) - (9 - \dot{m}_1)(3063) = 0$$

Solution of this equation gives:

$$\dot{m}_1 = 0.815\ kg/s$$

The mass flow rate $\dot{m}_2 = 8.185\ kg/s$.

Problem 2

Estimate the enthalpy of pure benzene in superheated vapor state at 100 °C and 1 atm pressure with respect to the following reference state: "$T_R = 80.1^oC$, $P_R = 1\ atm$, $h_R = 0$, saturated liquid benzene". Given that: $\Delta H^{vap} = 30{,}820\ kJ/kmol$ at normal boiling point of 80.1 °C, $C_{pG} = 96.3\ kJ/kmol.K$.

Solution

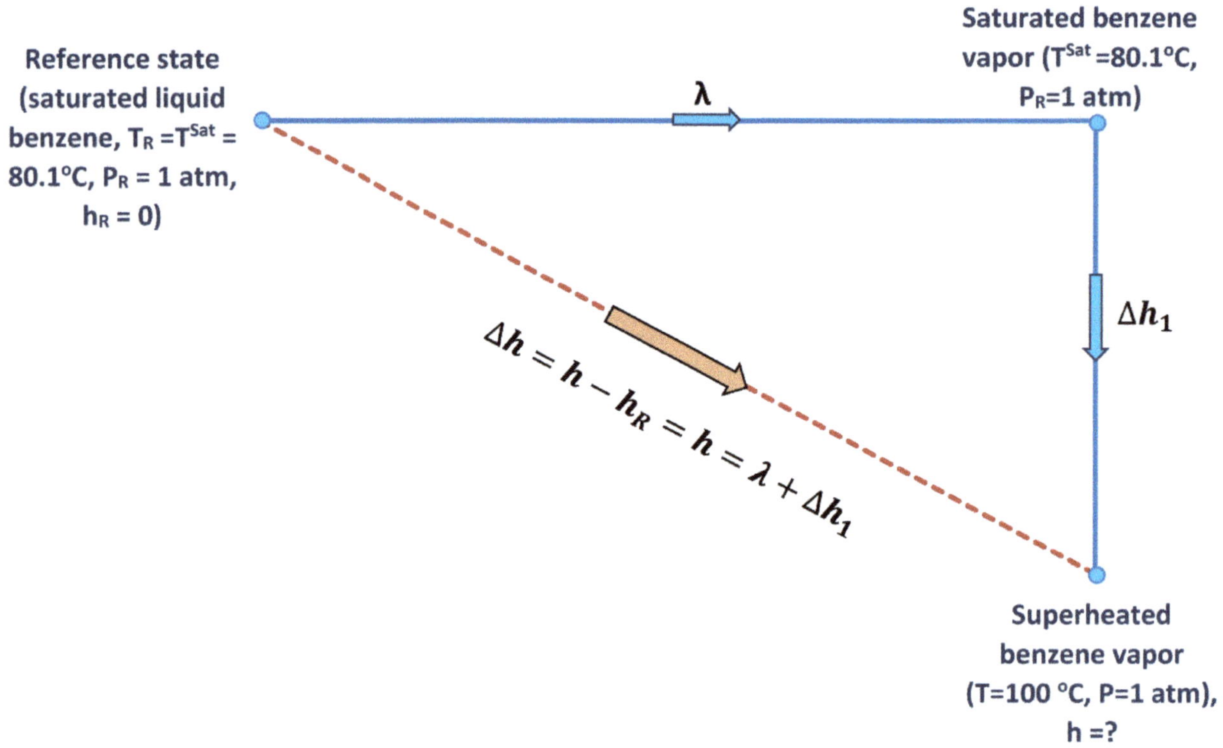

As enthalpy is a state property, the change in enthalpy is the same for different paths connecting the same initial and final states. Therefore,

$$\Delta h = \lambda + \Delta h_1$$

$$\Delta h = h - h_R = h$$

$$h = \lambda + C_{pG}(T - T^{Sat})$$

where λ is the latent heat of vaporization (ΔH^{vap}) and C_{pG} is constant pressure heat capacity of vapor (gas). Thus,

$$h = \lambda + C_{pG}(T - T^{Sat}) = 30820 + 96.3(100 - 80.1) = 32{,}736.37 \text{ kJ/kmol}$$

Chapter 8

Single-Stage Flash Distillation

Problem 1

A flash chamber operating at 1 atm is separating an equimolar mixture of benzene and toluene entering at flow rate of 1000 kmol/h. The feed mixture is 50 mole % benzene. If ψ=0.33, what are the product compositions and flow rates.
The equilibrium data for benzene-toluene system, in terms of mole fractions of benzene, at 1 atm are given below:

x	y
0	0
0.13	0.261
0.258	0.456
0.411	0.632
0.581	0.777
0.78	0.90
1	1

Solution

For ψ=0.33 and z_F =0.50, the equation for the operating line, Equation (8.3), is:

$$y_D = -2.03\, x_W + 1.515$$

The intersection of this operating line with the equilibrium curve gives the solution. Figure given below shows the graphical solution. The vapor distillate composition is $y_D \approx 0.646$ and the liquid composition is $x_W \approx 0.428$.

$$D = \psi F = 0.33 \times 1000 = 330 \text{ kmol/h}$$

$$W = (1 - \psi)F = 0.67 \times 1000 = 670 \text{ kmol/h}$$

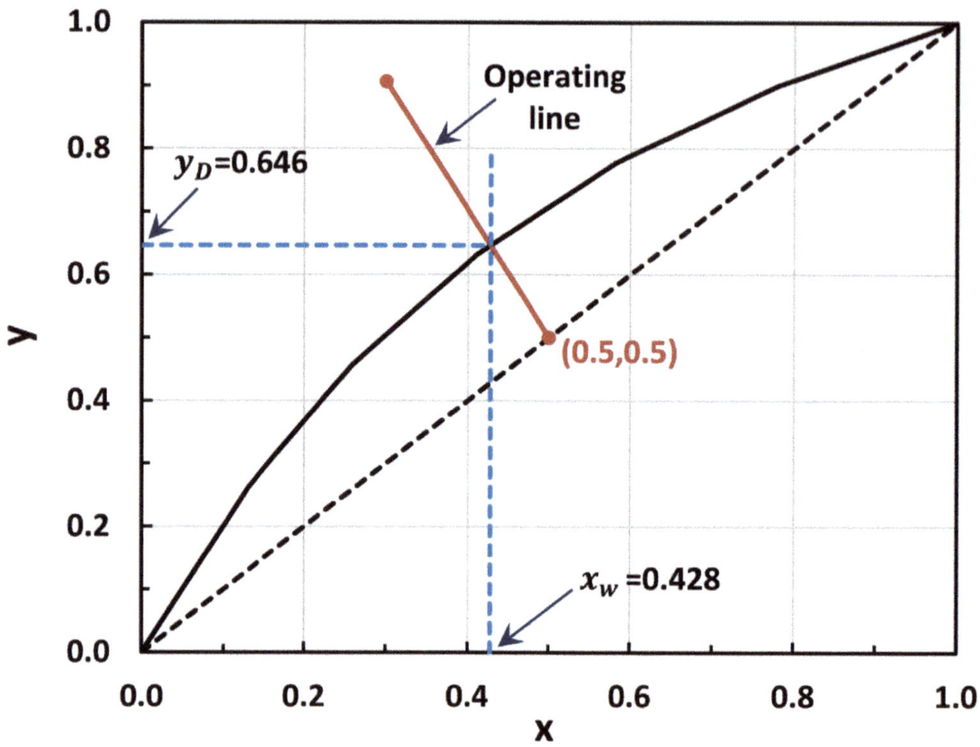

Problem 2

A 4-component feed mixture of species 1 to 4 with composition z_1=0.25, z_2 = 0.15, z_3=0.20 and z_4=0.40 undergoes flash distillation at 50 °C and 200 kPa. First verify that flashing will occur at the given conditions. What is the fraction of feed that goes to the vapor distillate product and what are the product compositions?
You can assume that Raoult's law is applicable. The vapor pressures of the pure species at 50 °C are as follows: $P_1^{Sat} = 1400\ kPa,\ P_2^{Sat} = 480\ kPa,\ P_3^{Sat} = 160\ kPa,\ P_4^{Sat} = 60\ kPa$

Solution

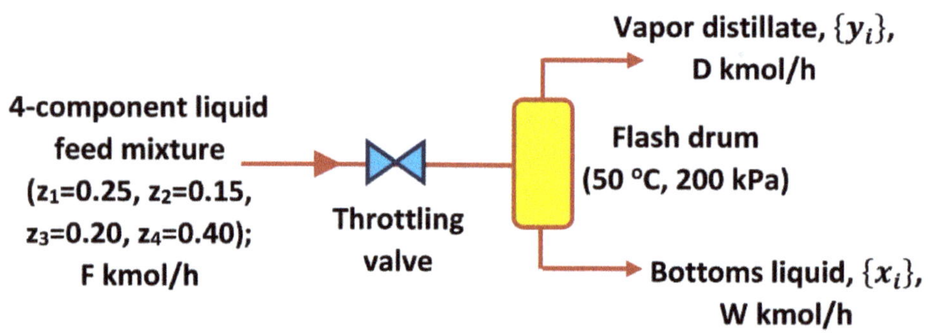

The problem deals with PT-flash calculation of multicomponent systems described in section 8.4. For flashing to occur, $P_{dew} < P < P_{bubble}$. The bubble and dew pressures are as follows:

$$P_{bubble} = 0.25(P_1^{Sat}) + 0.15(P_2^{Sat}) + 0.20(P_3^{Sat}) + 0.40(P_4^{Sat}) = 478\ kPa$$

$$P_{dew} = 1/[(0.25/P_1^{Sat}) + (0.15/P_2^{Sat}) + (0.20/P_3^{Sat}) + (0.40/P_4^{Sat})] = 118.94\ kPa$$

As the pressure in the chamber (200 kPa) falls in the range $P_{dew} < P < P_{bubble}$, flashing will occur at the given chamber conditions.

We now need to solve the Rachford-Rice equation, Equation (8.43), using the Newton-Raphson method, given that: z_1=0.25, z_2 = 0.15, z_3=0.20, z_4=0.40, $K_1 = P_1^{sat}/P = 7$, $K_2 = P_2^{sat}/P = 2.4$, $K_3 = P_3^{sat}/P = 0.80$, $K_4 = P_4^{sat}/P = 0.30$

The Rachford-Rice equation in the present case can be written as:

$$f(\psi) = \sum \frac{(K_i - 1)z_i}{1 + \psi(K_i - 1)} = 0$$

$$f(\psi) = \frac{1.5}{1 + \psi(6.0)} + \frac{0.21}{1 + \psi(1.4)} - \frac{0.04}{1 - \psi(0.2)} - \frac{0.28}{1 - \psi(0.7)} = 0$$

The derivative of $f(\psi)$ is as follows:

$$f'(\psi) = -\sum \frac{(K_i - 1)^2 z_i}{[1 + \psi(K_i - 1)]^2}$$

$$f'(\psi) = -\left[\frac{9}{(1 + \psi(6))^2} + \frac{0.294}{(1 + \psi(1.4))^2} + \frac{0.008}{(1 - \psi(0.2))^2} + \frac{0.196}{(1 - \psi(0.7))^2}\right]$$

Equation (8.48) gives the iteration scheme to solve for ψ as:

$$\psi_{k+1} = \psi_k - \frac{f(\psi_k)}{f'(\psi_k)}$$

The following table gives the iterations starting with the initial guess of ψ_o=0.50

ψ_k	ψ_{k+1}
0.50	0.520488
0.520488	0.520633
0.520633	0.520633
0.520633	0.520633

Thus the solution is ψ=0.520633≈0.5206.

The vapor composition follows from Equation (8.39):

$$y_i = \frac{K_i z_i}{1 + \psi(K_i - 1)}$$

The liquid composition follows from the equilibrium relation:

$$x_i = y_i / K_i$$

The values of y_i and x_i are summarized below.

species	y_i	x_i
1	0.424386	0.060627
2	0.208232	0.086763
3	0.178595	0.223244
4	0.188804	0.629346

Chapter 9

Multi-Stage Column Distillation

Problem 1

Feed entering the distillation column is saturated liquid at column conditions. It consists of a binary mixture of benzene and toluene. The mole fraction of benzene in the feed is 0.57. The feed flow rate is 29 kmol/h. The distillate product contains 95 mol% benzene and the bottoms product contains 3.0 mol% benzene. The reflux ratio is 3.25. Do the following:

a) Calculate the distillate product flow rate, D
b) Calculate the bottoms product flow rate, W
c) Determine the optimum number of equilibrium stages using the McCabe-Thiele method
d) Determine the optimum feed tray location
e) Plot Txy diagram
f) Plot composition profile of the column
g) Plot temperature profile of the column

Assume Raoult's law is applicable and the column pressure is 101.32 kPa. The vapor pressures of pure benzene and pure toluene are given by the following Antoine equations:

$$lnP_{benzene}^{Sat} = 13.7819 - \frac{2726.81}{t+217.572}$$

$$lnP_{toluene}^{Sat} = 13.932 - \frac{3056.96}{t+217.625}$$

where t is in °C and the vapor pressures are in kPa.

Solution

Refer to the column diagram given below.

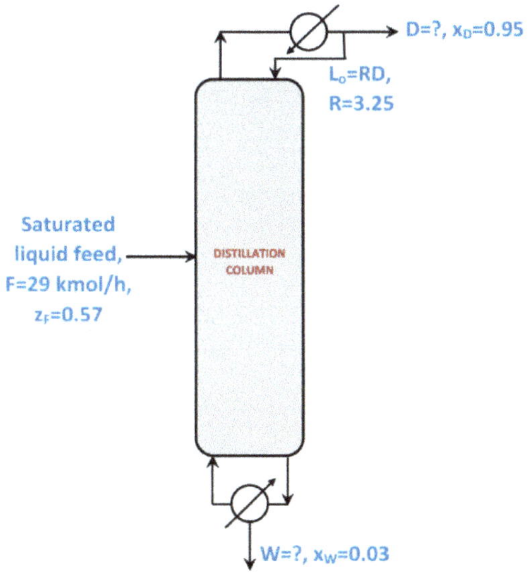

a)

$$\frac{D}{F} = \frac{z_F - x_W}{x_D - x_W} = \frac{0.57-0.03}{0.95-0.03} = 0.587$$

$$D = 0.587F = 0.587 \times 29 = 17.02\ kmol/h$$

b)

$$\frac{W}{F} = \frac{x_D - z_F}{x_D - x_W} = \frac{0.95-0.57}{0.95-0.03} = 0.413$$

$$W = 0.413F = 0.413 \times 29 = 11.98\ kmol/h$$

c) As the feed is saturated liquid, $q = 1$ and the slope of q-line is ∞. Thus q-line is a vertical line passing through the point (0.57, 0.57). The enriching section operating line passes through the point (0.95, 0.95) and has a y-intercept of $x_D/(1+R) = 0.95/(1+3.25) = 0.2235$. Before we can estimate the optimum number of equilibrium stages required using the McCabe-Thiele graphical method, we need to generate the equilibrium data.

The boiling points of pure species 1 (benzene) and 2 (toluene) at 101.32 kPa are calculated from the given Antoine equations as: $T_1^{Sat} = 80^oC$ and $T_2^{Sat} = 110.60^oC$. Table given below shows at different temperatures the values of vapor pressures of pure species and mole fractions of more volatile component (species 1-benzene) in the vapor and liquid phases at equilibrium with each other. The Antoine equations are used to calculate the vapor pressures. Equations (6.44) and (6.45) of Chapter 6 are used to calculate the mole fractions of more volatile component in liquid and vapor phases at equilibrium. The T-x-y and y-x diagrams for the system at 101.32 kPa are developed using the data shown in the table. Note that x and y refer to mole fractions of more volatile species (species 1).

T (°C)	$P_1^{Sat}(kPa)$	$P_2^{Sat}(kPa)$	x	y
80.0	101.32	38.89	1.00	1.00
82.0	107.72	41.65	0.90	0.96
84.0	114.42	44.57	0.81	0.92
86.0	121.45	47.65	0.73	0.87
88.0	128.80	50.89	0.65	0.82
90.0	136.50	54.31	0.57	0.77
92.0	144.54	57.91	0.50	0.71
94.0	152.95	61.70	0.43	0.66
96.0	161.73	65.68	0.37	0.59
98.0	170.90	69.87	0.31	0.53
100.0	180.45	74.26	0.25	0.45
102.0	190.42	78.87	0.20	0.38
104.0	200.79	83.70	0.15	0.30
106.0	211.60	88.77	0.10	0.21
108.0	222.84	94.07	0.06	0.12
110.6	238.13	101.32	0.00	0.00

The following figure shows the plots of the operating lines and stepping off the stages using the McCabe-Thiele method. The optimum number of equilibrium stages required (including partial reboiler) is 10.

d) The optimum feed tray location is 4 (as shown in the figure below).

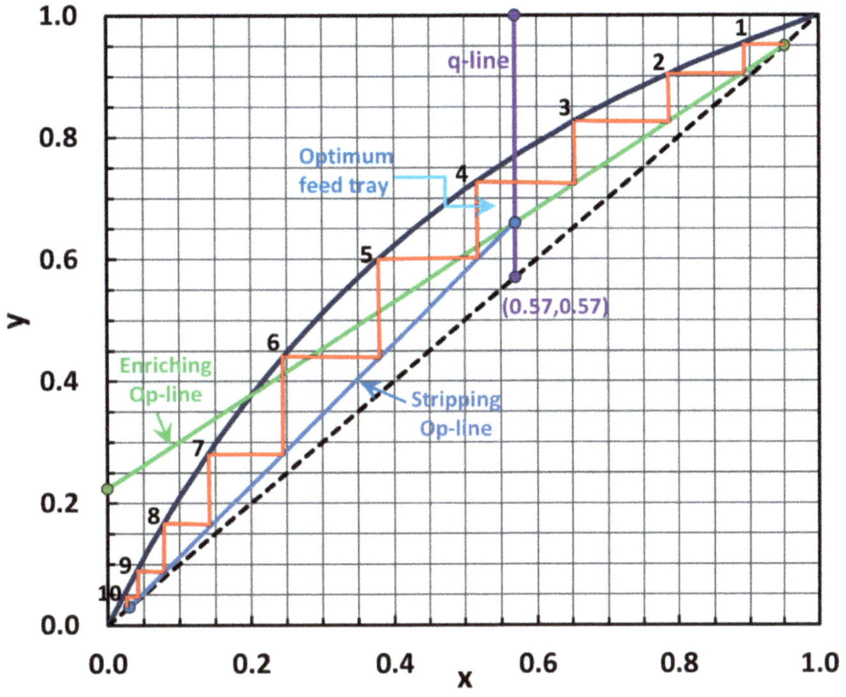

e) The Txy diagram is shown below.

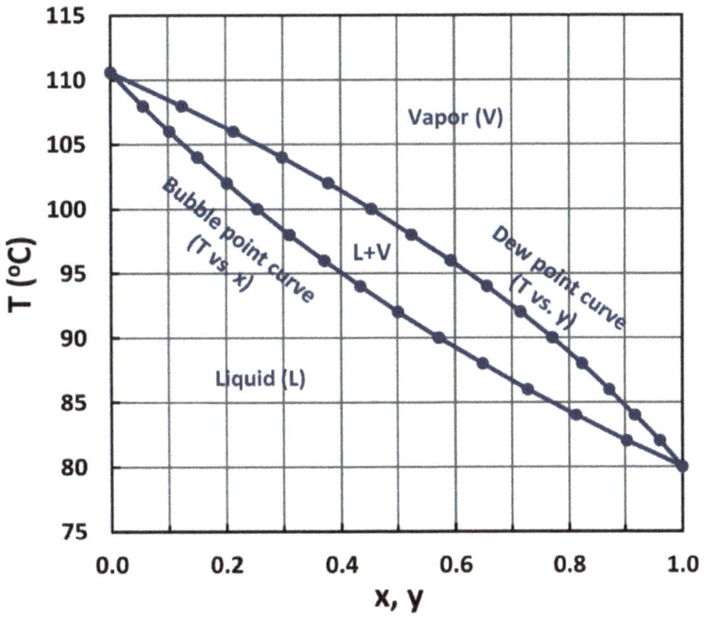

f) Composition profiles of the column

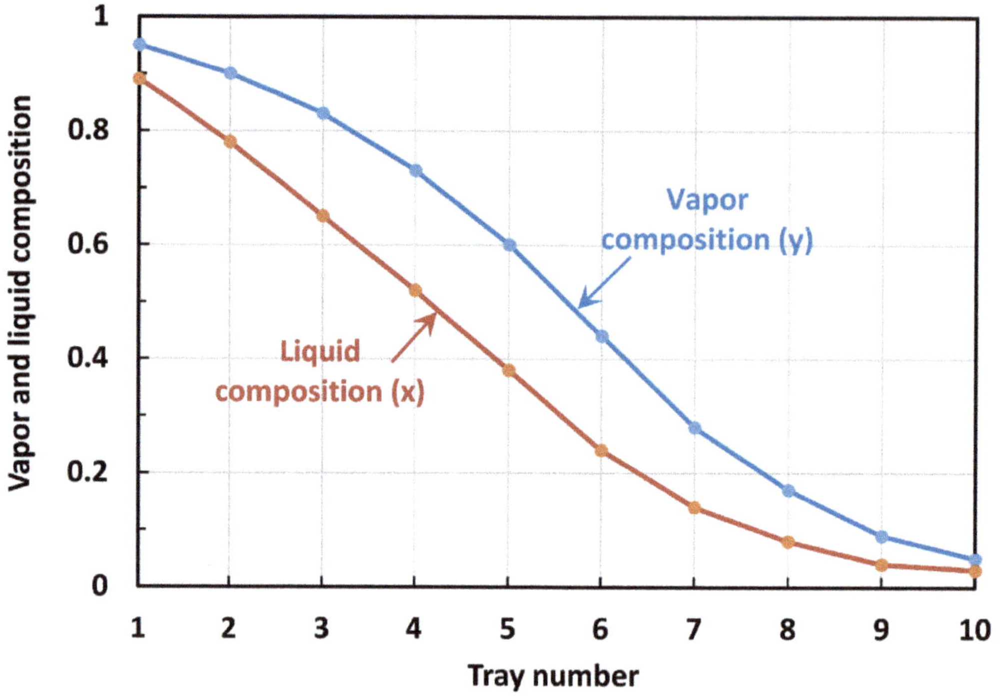

g) Temperature profile of the column

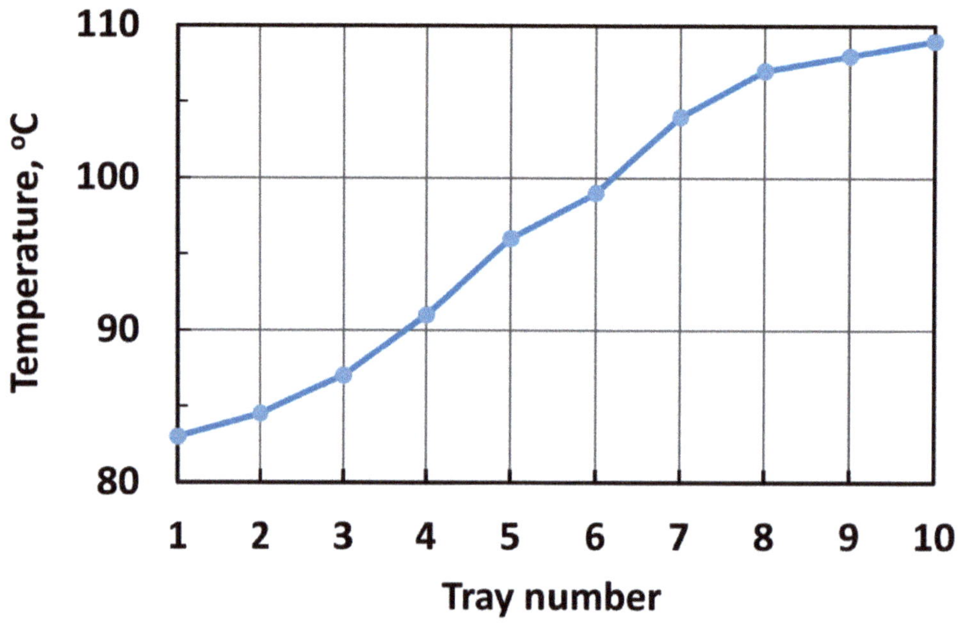

Problem 2

Determine graphically the minimum number of equilibrium stages required to separate a binary feed mixture of benzene and toluene containing 55 mol% benzene into a distillate product of composition 98 mol% benzene and a bottoms product of composition 5 mol% benzene.

Assume Raoult's law is applicable and the column pressure is 101.32 kPa. The vapor pressures of pure benzene and pure toluene are given by the following Antoine equations:

$$lnP_{benzene}^{Sat} = 13.7819 - \frac{2726.81}{t+217.572}$$

$$lnP_{toluene}^{Sat} = 13.932 - \frac{3056.96}{t+217.625}$$

where t is in °C and the vapor pressures are in kPa.

Solution

The equilibrium y-x curve is generated as described in problem 1 solution above. At total reflux, the operating lines coincide with y=x line. The minimum number of stages required (including partial reboiler) is about 8 as shown in the figure below.

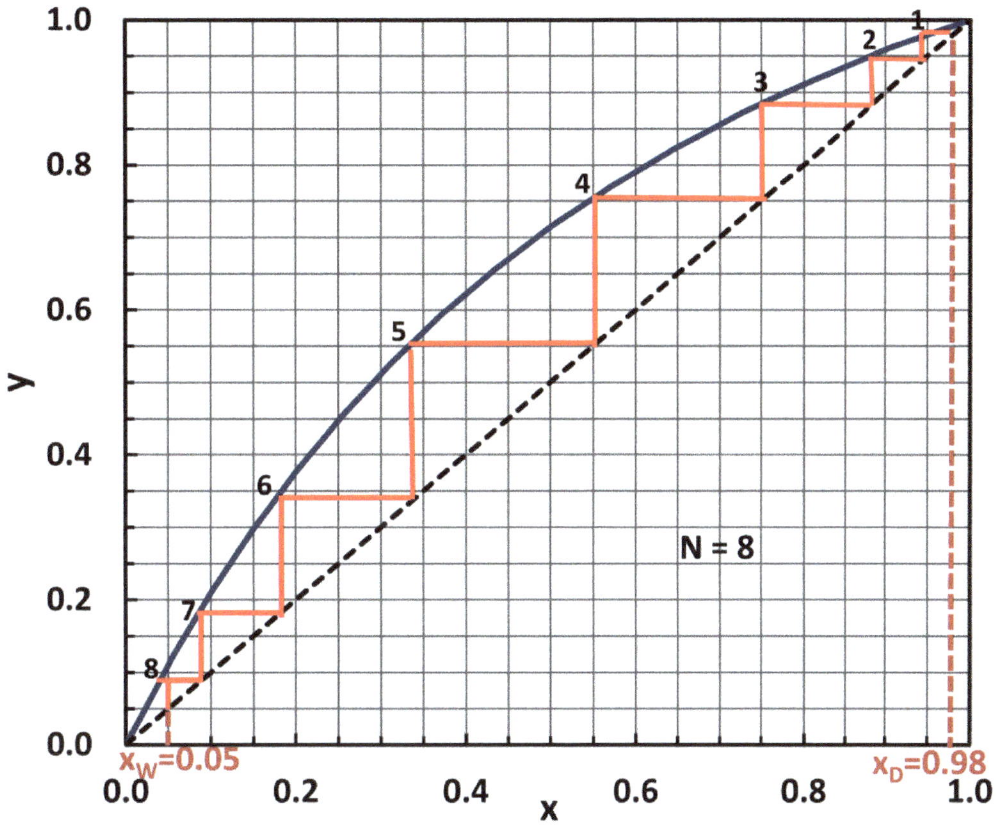

Problem 3

Feed entering the distillation column is a two-phase vapor-liquid mixture such that:

$$\frac{\bar{L} - L}{F} = 0.81$$

The feed consists of two components, namely ethanol and water. The overall composition of the feed is 50 mol% ethanol and the total feed flow rate is 300 kmol/h. The distillate product contains 78 mol% ethanol and the bottoms product contains 6 mol% ethanol. The reflux ratio is 2.90. The column is operating at 1

atm pressure with direct steam injection into the bottom of the column (instead of using a reboiler). The steam injected into the column is saturated steam at 1 atm pressure. Determine the number of equilibrium stages required for the given separation. What is the optimum feed tray location?
The equilibrium data for ethanol-water, in terms of mole fraction of more volatile component ethanol, at column pressure (1 atm) is as follows:

x	y
0	0
0.019	0.17
0.0721	0.3891
0.0966	0.4375
0.1238	0.4704
0.1661	0.5089
0.2377	0.5445
0.2608	0.558
0.3273	0.5826
0.3965	0.6122
0.5079	0.6564
0.5198	0.6599
0.5732	0.6841
0.6763	0.7385
0.7472	0.7815
0.8943	0.8943
1	1

Solution

Refer to the schematic diagram of the column shown below.

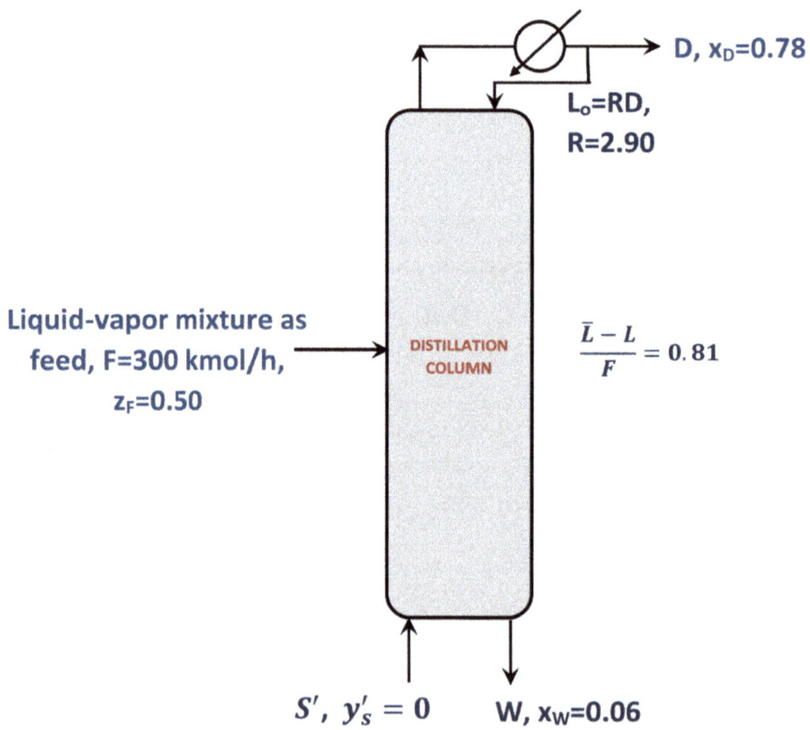

The enriching section operating line is as follows:

$$y_{n+1} = \left(\frac{R}{1+R}\right) x_n + \left(\frac{x_D}{1+R}\right) = \left(\frac{2.90}{1+2.90}\right) x_n + \left(\frac{0.78}{1+2.90}\right) = 0.7436x_n + 0.20$$

This line passes through the point (0.78, 0.78) and has a y-intercept of 0.20.

The feed is a two-phase vapor-liquid mixture consisting of 81 mol% liquid and 19 mol% vapor. The feed quality q factor is:

$$q = \frac{\bar{L} - L}{F} = 0.81$$

The equation of feed line, Equation (9.38) can now be written as:

$$y = -\left(\frac{q}{1-q}\right) x + \left(\frac{z_F}{1-q}\right) = -\left(\frac{0.81}{1-0.81}\right) x + \left(\frac{0.50}{1-0.81}\right) = -4.263x + 2.632$$

Thus the q-line (feed line) passes through the points (0.50, 0.50) and (0.383, 1.0).

The stripping section operating line, Equation (9.78), is:

$$y = \left(\frac{W}{S'}\right)(x - x_W) = \left(\frac{\bar{L}}{\bar{V}}\right)(x - x_W)$$

This line passes through the point $(x_w, 0)$ and the point of intersection of enriching section operating line and q-line. Note that $x_w = 0.06$.

The figure given below shows the plots of operating lines and q-line. The stepping off stages is also shown.

The total number of stages required is 7. The optimum feed location is stage number 6.

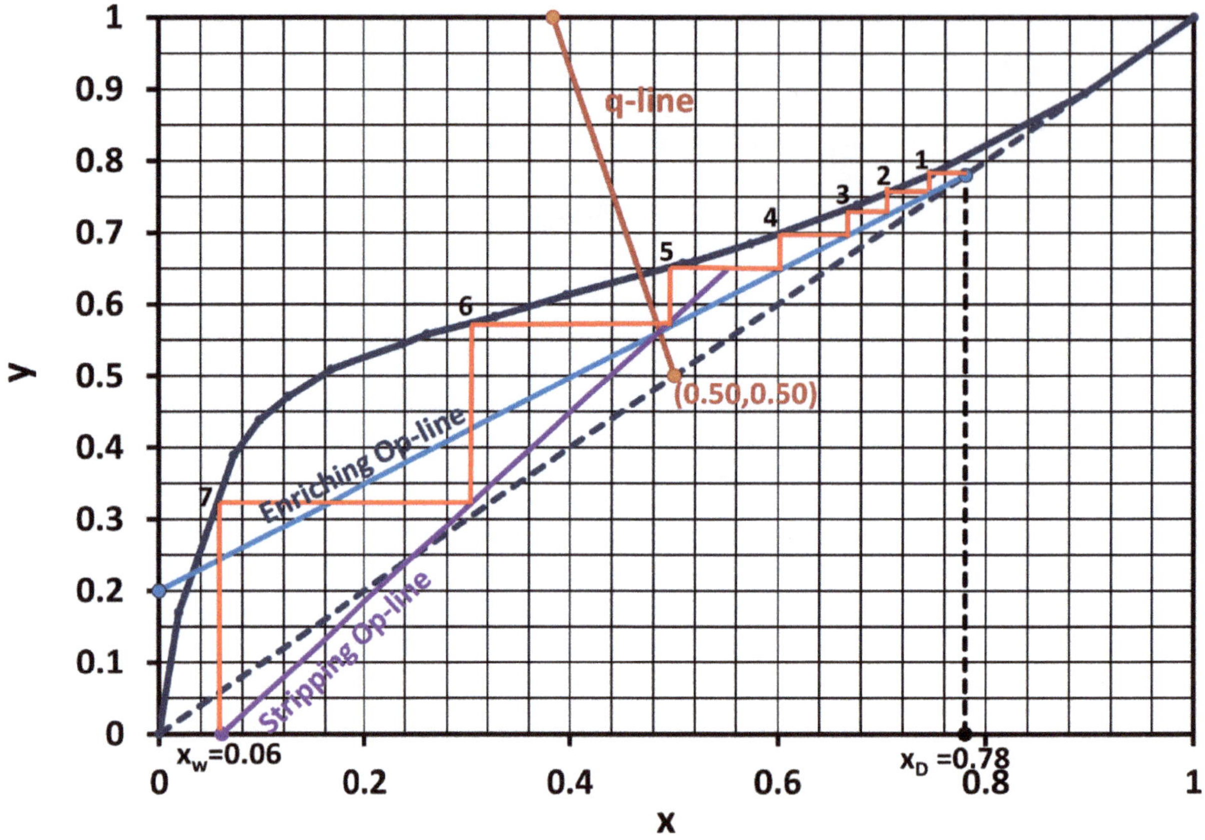

Problem 4

A multistage distillation column is operated at 1 atm pressure to separate acetone-ethanol mixture. The feed to the column is a saturated liquid consisting of 58 mol% acetone and has a flow rate of 98 kmol/h. The liquid side product is withdrawn from second stage (counting from the top of the column) at a flow rate of $S = 13$ kmol/h. The reflux ratio (R) is 3.0. The desired distillate and bottoms product compositions are: 89 mol% acetone and 9 mol% acetone, respectively.

Do the following:

a) Determine the mole fraction of acetone in the side stream x_S
b) Determine the optimum number of equilibrium stages using the McCabe-Thiele method
c) Determine the optimum feed tray location

The equilibrium data for acetone-ethanol, in terms of mole fraction of more volatile component acetone, at column pressure (1 atm) is given in tutorial problem 6.

Solution

Refer to the schematic diagram of the column shown below.

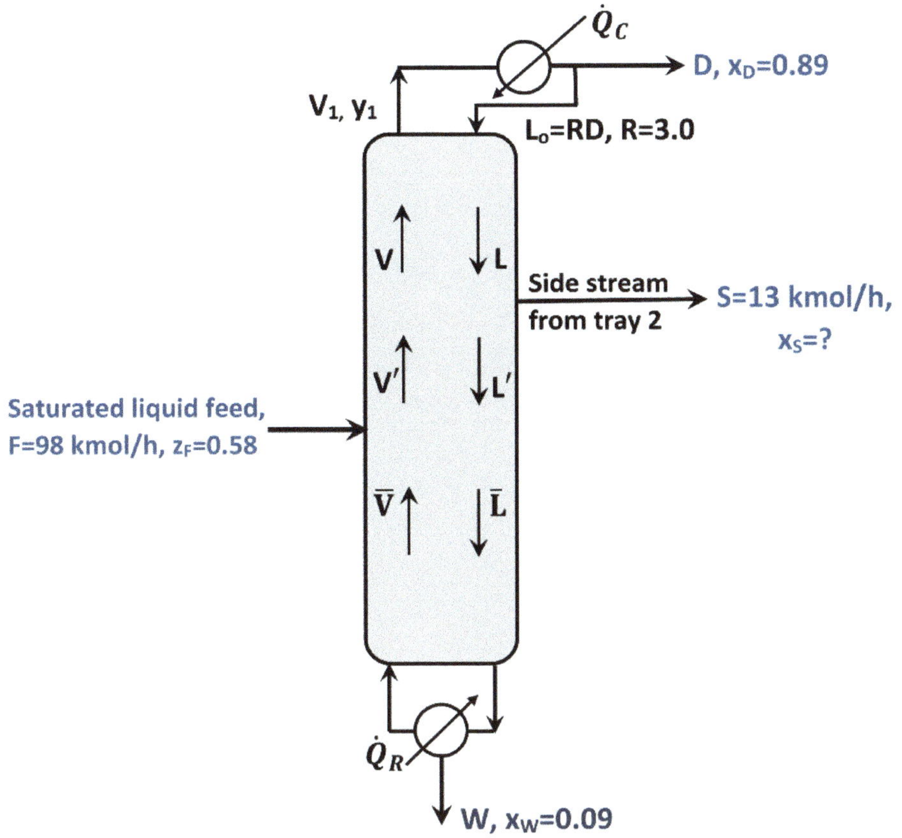

a) The mole fraction of acetone in the side stream x_S :

In order to estimate the side stream x_S, we are required to plot the enriching section operating line and step off two stages as the liquid side product is withdrawn from the second stage (counting from the top of the column). The figure below shows the stepping off two stages. According to the graph, the composition of the liquid side stream is $x_S = 0.80$. Note that the enriching section operating line passes through the point (0.89, 0.89) and has a y-intercept of 0.2225.

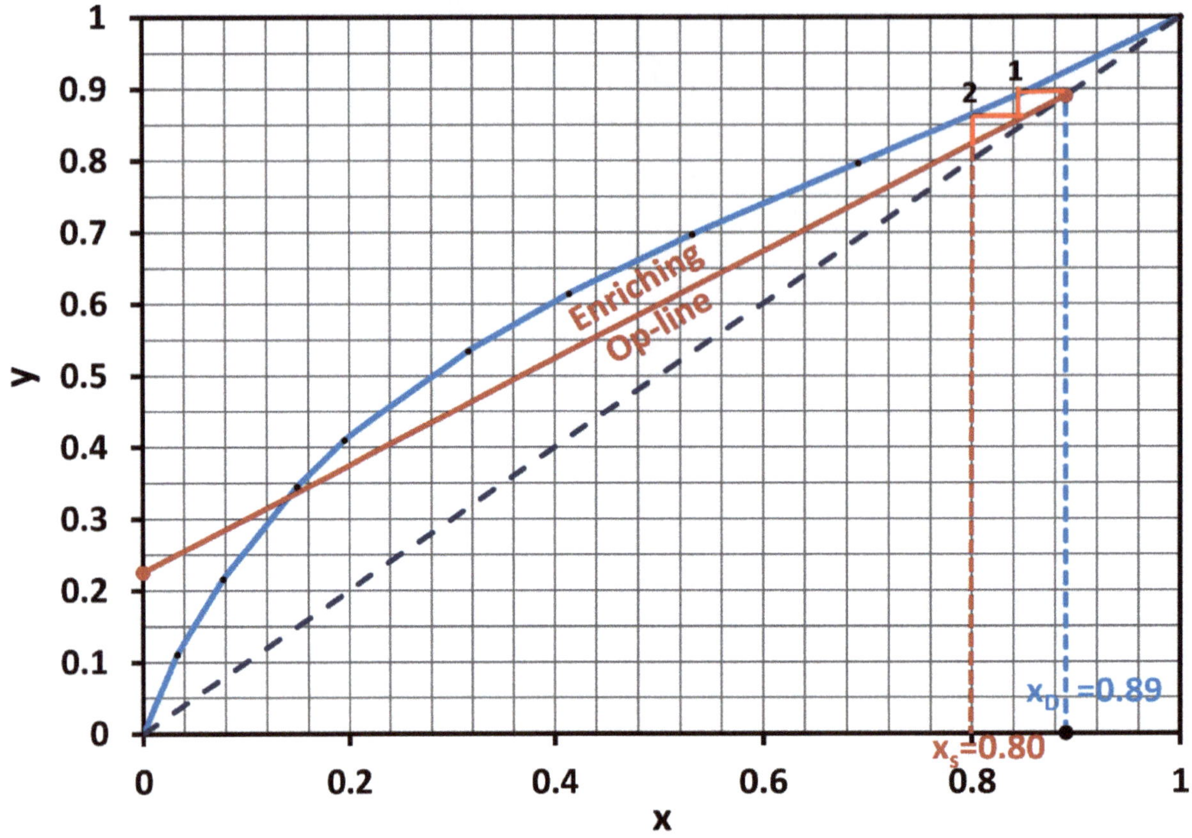

b) The optimum number of equilibrium stages using the McCabe-Thiele method:

Let us first calculate the distillate and bottoms product flow rate.

The distillate flow rate D is given by Equation (9.67) as:

$$D = \frac{(z_F - x_W)F - (x_S - x_W)S}{x_D - x_W} = \frac{(0.58-0.09)(98)-(0.80-0.09)(13)}{0.89-0.09} = 48.4875\ kmol/h$$

The bottoms product flow rate W is given by Equation (9.68) as:

$$W = \frac{(x_D - z_F)F - (x_D - x_S)S}{x_D - x_W} = \frac{(0.89-0.58)(98)-(0.89-0.80)(13)}{0.89-0.09} = 36.5125\ kmol/h$$

Note that the overall material balance gives the same value of W:

$$W = F - D - S = 98 - 48.4875 - 13 = 36.5125\ kmol/h$$

The side stream q-line passes through the point (0.80, 0.80). It is a vertical line as the side stream withdrawn is saturated liquid.

The feed q-line passes through the point (0.58, 0.58). It is a vertical line as the feed is saturated liquid.

The enriching section operating line is:

$$y_{n+1} = \left(\frac{R}{1+R}\right)x_n + \left(\frac{x_D}{1+R}\right) = \left(\frac{3.0}{1+3.0}\right)x_n + \left(\frac{0.89}{1+3.0}\right) = 0.75x_n + 0.2225$$

This line passes through the point (0.89, 0.89) and has a y-intercept of 0.2225. It intersects the side q-line at (0.80, 0.8225).

The middle section operating line is given by Equation (9.69) as:

$$y = \left(\frac{L'}{V'}\right)x + \left(\frac{x_D\,D + x_S S}{V'}\right)$$

$$L' = RD - S = 3 \times 48.4875 - 13 = 132.4625\ kmol/h$$

$$V' = (1 + R)D = 4 \times 48.4875 = 193.95\ kmol/h$$

Thus, the middle section operating line is:

$$y = 0.683x + 0.276$$

The middle section line passes through the point (0.80, 0.822) and intersects the feed q-line at (0.58, 0.672). Finally the stripping section operating line is drawn. It passes through the point (0.09, 0.09) and the intersection point of feed q-line and middle section line at (0.58, 0.672).

All the operating and q-lines are drawn as shown in the figure given below. The stepping off stages, as shown in the figure, gives the optimum total number of stages to be slightly more than 9 (including partial reboiler).

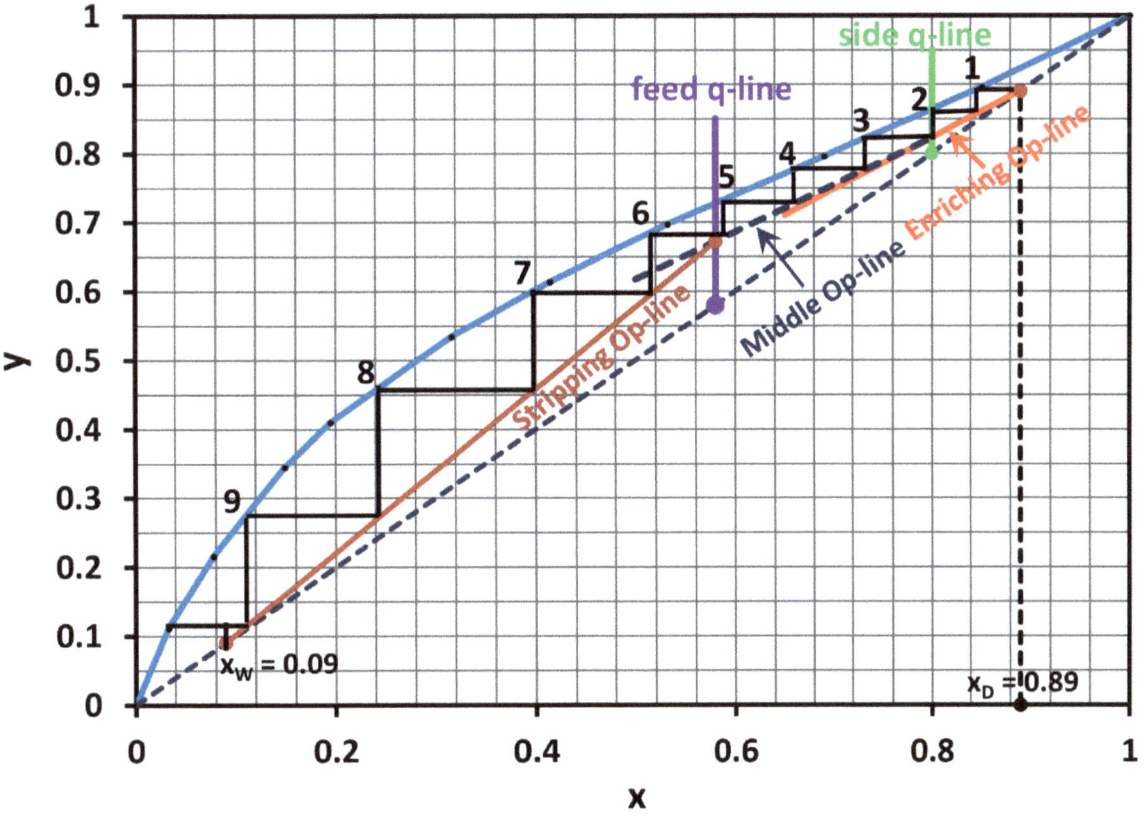

c) The optimum feed tray location:
The optimum feed tray is 6 counting from top of the column.

Problem 5

A saturated liquid mixture of benzene and toluene containing 68 mol% benzene is fed to a stripping column at a flow rate of 388 kmol/h. The column is operating at 101.32 kPa. The bottoms product contains 9 mol% benzene and has a flow rate of 58 kmol/h.
Calculate the composition and flow rate of overhead vapor product. How many equilibrium stages are required?
Assume Raoult's law is applicable. The vapor pressures of pure benzene and pure toluene are given by the following Antoine equations:

$$lnP^{Sat}_{benzene} = 13.7819 - \frac{2726.81}{t+217.572}$$

$$lnP^{Sat}_{toluene} = 13.932 - \frac{3056.96}{t+217.625}$$

where t is in °C and the vapor pressures are in kPa.

Solution

Refer to the schematic diagram of the column shown below.

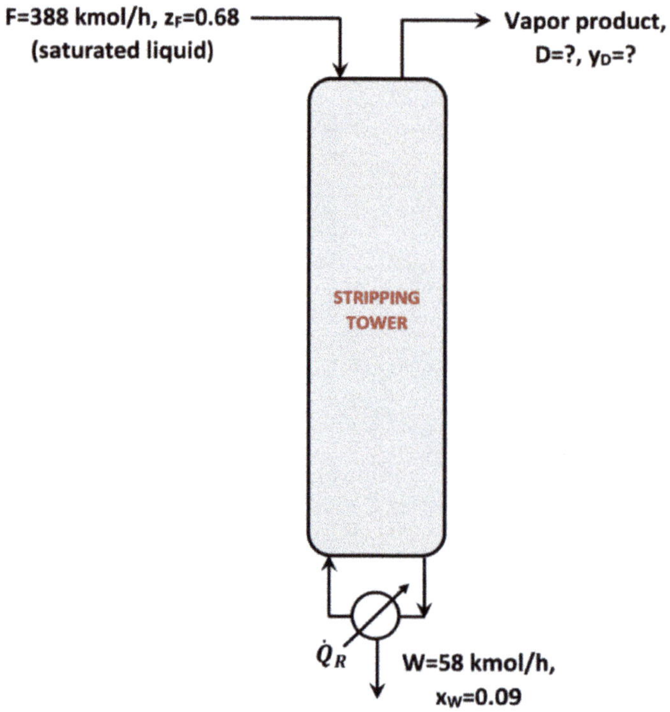

Overall material balance:

$$F = D + W$$

$$D = F - W = 388 - 58 = 330 \text{ kmol/h}$$

$$\bar{L} = F = 388 \text{ kmol/h}$$

$$\bar{V} = D = 330 \text{ kmol/h}$$

Stripping section operating line:

$$y_{m+1} = \left(\frac{\bar{L}}{\bar{V}}\right) x_m - \left(\frac{W}{\bar{V}}\right) x_w$$

$$y_{m+1} = \left(\frac{388}{330}\right) x_m - \left(\frac{58}{330}\right) x_w$$

$$y_{m+1} = 1.1758 x_m - 0.0158$$

The stripping section operating line passes through (0.09, 0.09) and has a y-intercept of -0.0158.
The equation of feed line, Equation (9.38, is:

$$y = -\left(\frac{q}{1-q}\right) x + \left(\frac{z_F}{1-q}\right)$$

As the feed is saturated liquid, $q = 1$ and the slope of the operating line is $-\infty$. Thus the feed line (q-line) is a vertical line passing through the point (0.68, 0.68).

The intersection of the stripping operating line and q-line gives the composition of the overhead vapor product.

The equilibrium y-x curve is generated using the Antoine equations as described in the solutions of problems 1 and 2 above. The figure given below shows the plots of operating lines and q-line. The stepping off stages is also shown. Note that the intersection of the stripping operating line and q-line gives $y_D = 0.785$ as the composition of the overhead vapor product. Thus the enriching section operating line is $y = y_D = 0.785$.

The stepping off stages gives the total number of stages required (including partial reboiler) is about 5.3.

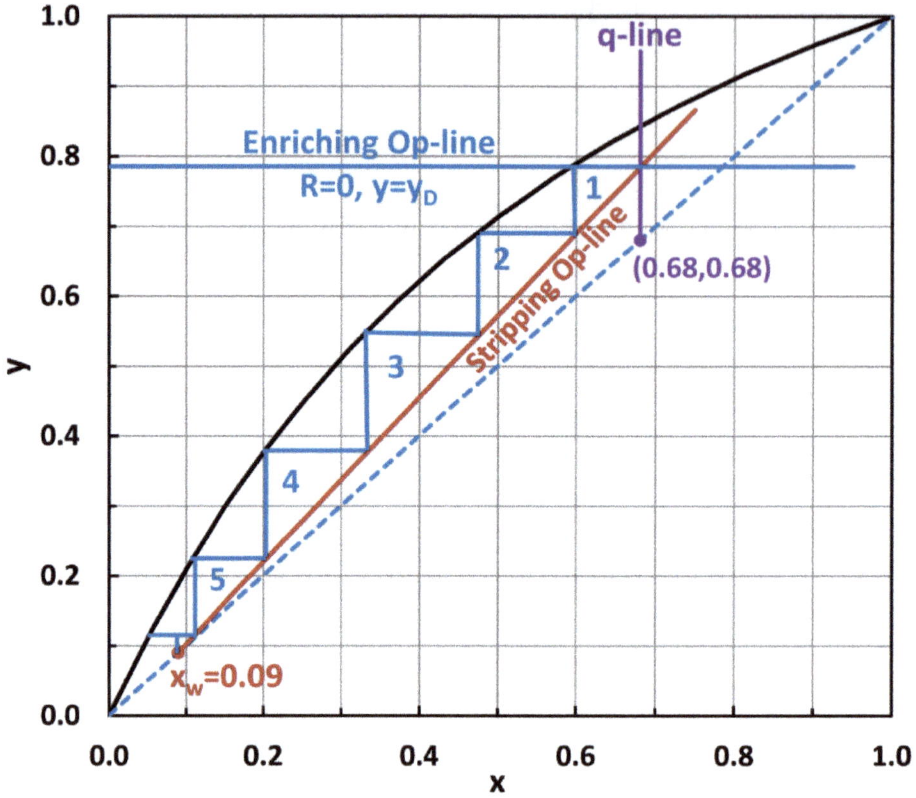
1.0
0.8
0.6
0.4
0.2
0.0
y
0.0
0.2
0.4
0.6
0.8
1.0
x
q-line
Enriching Op-line
R=0, y=y_D
1
2
3
4
5
Stripping Op-line
(0.68,0.68)
x_w=0.09

Chapter 10

Approximate Analytical Method of Designing Distillation Columns

Problem 1

A thin cylindrical vessel of diameter 1.15 m is subjected to an internal pressure of 2.1 N/mm^2. Estimate the minimum wall thickness if the longitudinal stress does not exceed 30 MPa. What is the minimum wall thickness if the hoop stress does not exceed 45 MPa?

Solution

Using Equation (10.51), the minimum wall thickness based on longitudinal stress can be calculated as:

$$t_s = P_i D_i / 4\sigma_p = 2.1 \times 1150/(4 \times 30) = 20.125\ mm$$

Using Equation (10.56), the minimum wall thickness based on hoop stress can be calculated as:

$$t_s = P_i D_i / 2\sigma_h = 2.1 \times 1150/(2 \times 45) = 26.83\ mm$$

Thus the minimum wall thickness must be 26.83 mm in order to withstand the maximum longitudinal and hoop stresses.

Problem 2

A distillation column consisting of sieve plates is separating a hydrocarbon mixture feed. The mass flow rate of vapor at the bottom of the column is 45 lb$_m$/s. The ratio of liquid mass flow rate to vapor mass flow rate is 0.85. The densities of vapor and liquid are 0.2 lb$_m$/ft^3 and 40 lb$_m$/ft^3, respectively. Assuming that the column operates at 80% of the flooding velocity, estimate the column diameter. The liquid surface tension is 15 dynes/cm and the tray spacing is 24 inches.

Solution

Given: $\dot{m}_V = 45\ lb_m/s$, $\dot{m}_L/\dot{m}_V = 0.85$, $\rho_V = 0.2\ lb_m/ft^3$, $\rho_L = 40\ lb_m/ft^3$, $\gamma = 15\ dynes/cm$, $h_p = 24\ inches$, $U = 0.80(U_f)$

The flow parameter F_{LV} is:

$$F_{LV} = \left(\frac{\dot{m}_L}{\dot{m}_V}\right)\sqrt{\frac{\rho_V}{\rho_L}} = 0.85\sqrt{\frac{0.2}{40}} = 0.0601$$

Using Equation (10.45), we can calculate C

$$log_{10}C = -0.94506 - 0.70234[log_{10}(0.0601)] - 0.22618[log_{10}(0.0601)]^2$$

This equation gives a C value of 0.3761 ft/s. The flooding velocity is calculated from Equation (10.44):

$$U_f = 0.3761\left(\frac{15}{20}\right)^{0.2}\sqrt{\left(\frac{40-0.20}{0.20}\right)} = 5.01\ ft/s$$

Thus, the vapor velocity in the column is 0.80(5.01)=4.008 ft/s.

The column diameter can be estimated from Equation (10.49):

$$D = 2\sqrt{\frac{\dot{m}_V}{0.80\pi U_f\left(1-\frac{A_d}{A}\right)\rho_V}} = 2\sqrt{\frac{45}{0.80\pi(5.01)(1-0.1)(0.20)}} = 8.912\ ft$$

Note that the area ratio A_d/A is 0.1 as $F_{LV} \leq 0.1$.

Chapter 11

Ponchon-Savarit Graphical Method of Designing Distillation Columns

Problem 1

A multistage distillation column is to be operated at 1 atm pressure to separate ethanol-water mixture. The feed is saturated liquid containing 25 wt% ethanol. The flow rate of feed is 7,000 kg/h. The desired distillate composition is 82 wt.% ethanol and the bottoms composition is 3 wt.% ethanol. The column uses a total condenser and a partial reboiler.

Using the Ponchon-Savarit method, do the following:

a) Estimate the minimum reflux ratio R_{min}

b) Estimate the number of stages required at total reflux

c) Calculate condenser and reboiler heat duties at a reflux ratio of $1.3R_{min}$.

d) Estimate the number of equilibrium stages at $R = 1.3R_{min}$ with optimum feed location.

The equilibrium data for mixtures of ethanol and water at 1 atm pressure are given below.

Composition, mass fraction of ethanol		Enthalpy-data, kJ/kg	
Liquid (x)	**Vapor (y)**	**Liquid, h_L**	**Vapor, H_V**
0	0	418.9	2675
0.02	0.192	-	-
0.04	0.325	-	-
0.05	0.377	393.8	2593.5
0.10	0.527	371.7	2493.5
0.20	0.656	335.6	2355.1
0.30	0.713	314	2193.4
0.50	0.771	285.9	1870.1
0.60	0.794	273.3	1707.3
0.70	0.822	258.4	1544.5
0.80	0.858	241.4	1386.3
0.82	0.868	-	-
0.86	0.888	-	-
0.90	0.912	225	1223.5
1.0	0.978	207	1064.1

Solution

Refer to column schematic, y $vs.$ x equilibrium curve, and enthalpy-composition diagram shown below.

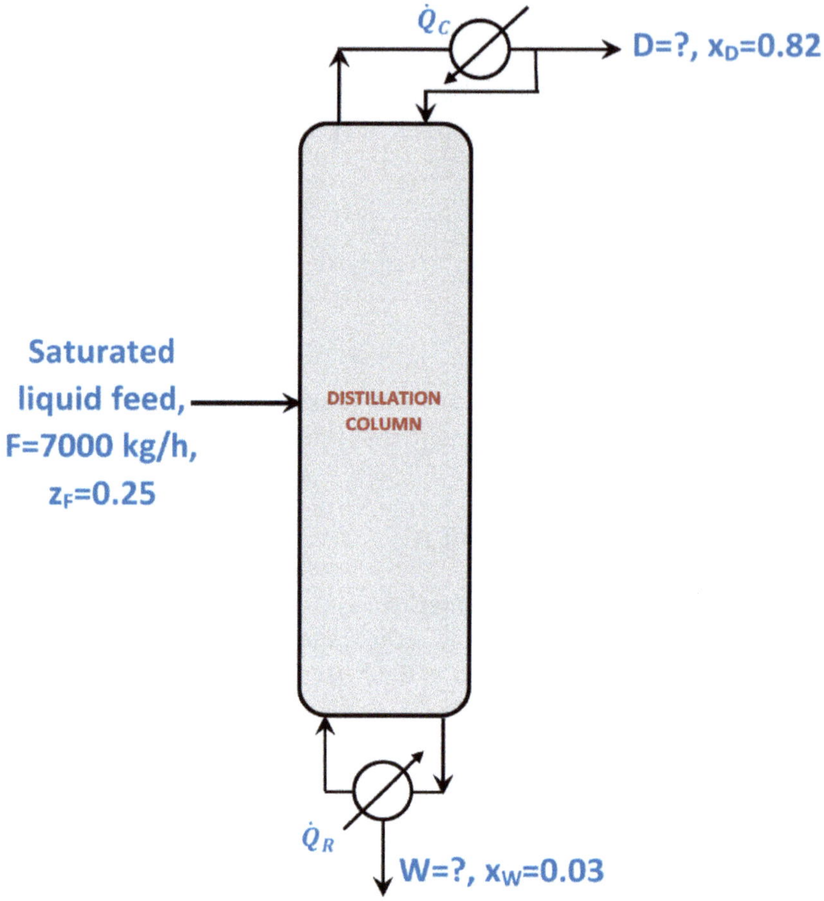
$\dot{Q}_C$
D=?, x_D=0.82
Saturated liquid feed, F=7000 kg/h, z_F=0.25
DISTILLATION COLUMN
$\dot{Q}_R$
W=?, x_W=0.03

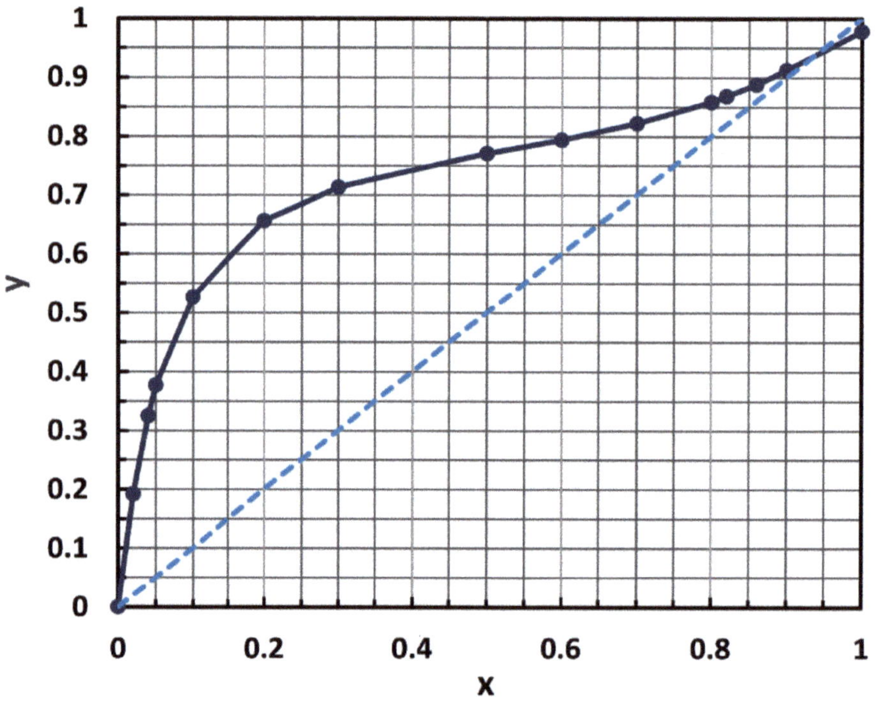
1
0.9
0.8
0.7
0.6
0.5
0.4
0.3
0.2
0.1
0
y
0
0.2
0.4
0.6
0.8
1
x

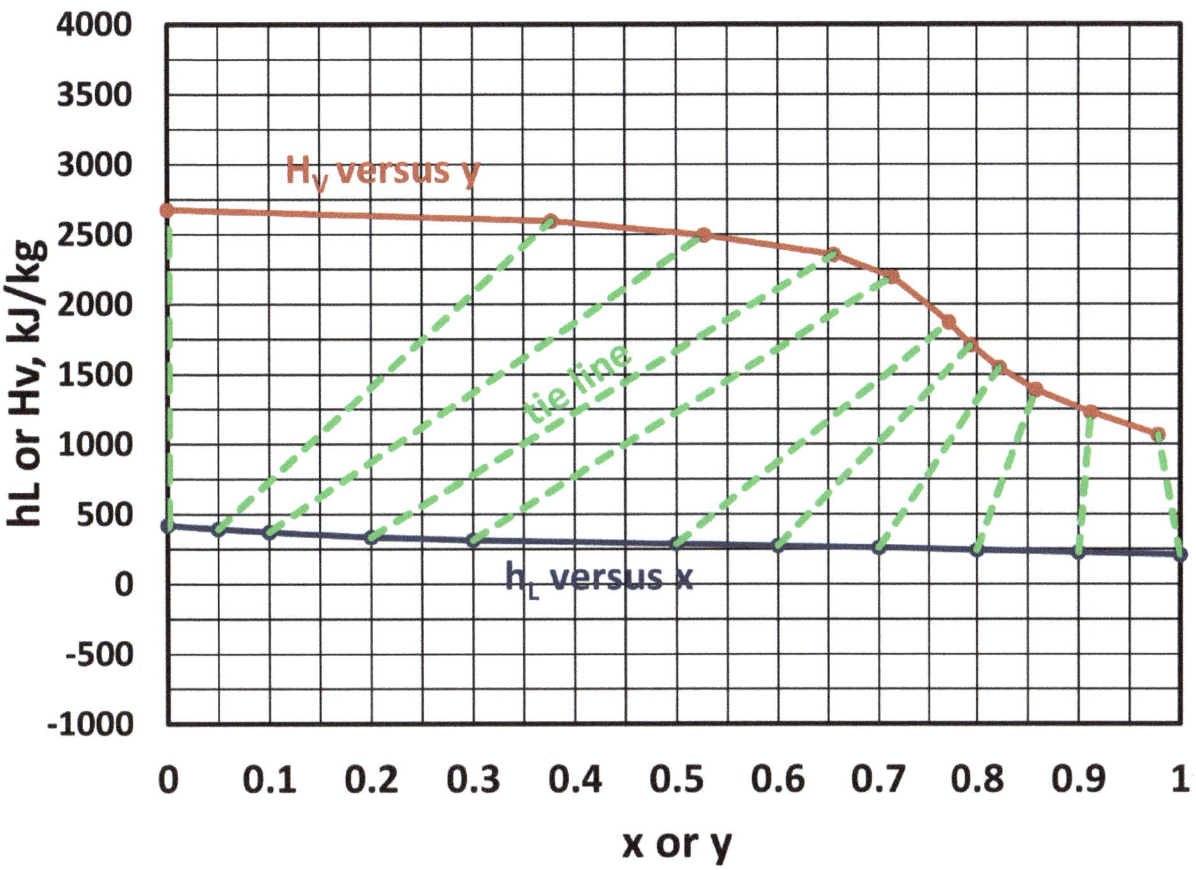

Note that ethanol-water system forms azeotrope at 96.6 wt.% ethanol (3.4 wt.% water) at 1 atm pressure. The equilibrium data shown in the $y\ vs.x$ equilibrium curve and enthalpy-composition diagram may not be accurate above the azeotrope composition ($x, y \geq 0.966$).

a) Estimation of the minimum reflux ratio R_{min}

As discussed in section 11.5.6, minimum reflux ratio is determined by extending the tie-line passing through the feed point $P_F(z_F, h_F)$ to the intersections with the two vertical lines corresponding to the product compositions ($x = x_D$ vertical line and $x = x_W$ vertical line). The extension of tie line passing through the feed point is shown in the figure given below. From the upper difference point enthalpy estimated graphically, the minimum reflux is calculated using Equation (11.42):

$$R_{min} = \left(h_{diff} - H_D\right)/\lambda_D$$

where $h_{diff} = 2800\ kJ/kg$, $H_D = 1550\ kJ/kg$, and $\lambda_D = H_D - h_D = 1550 - 240 = 1310\ kJ/kg$.

Thus,

$$R_{min} = (2800 - 1550)/1310 = 0.954$$

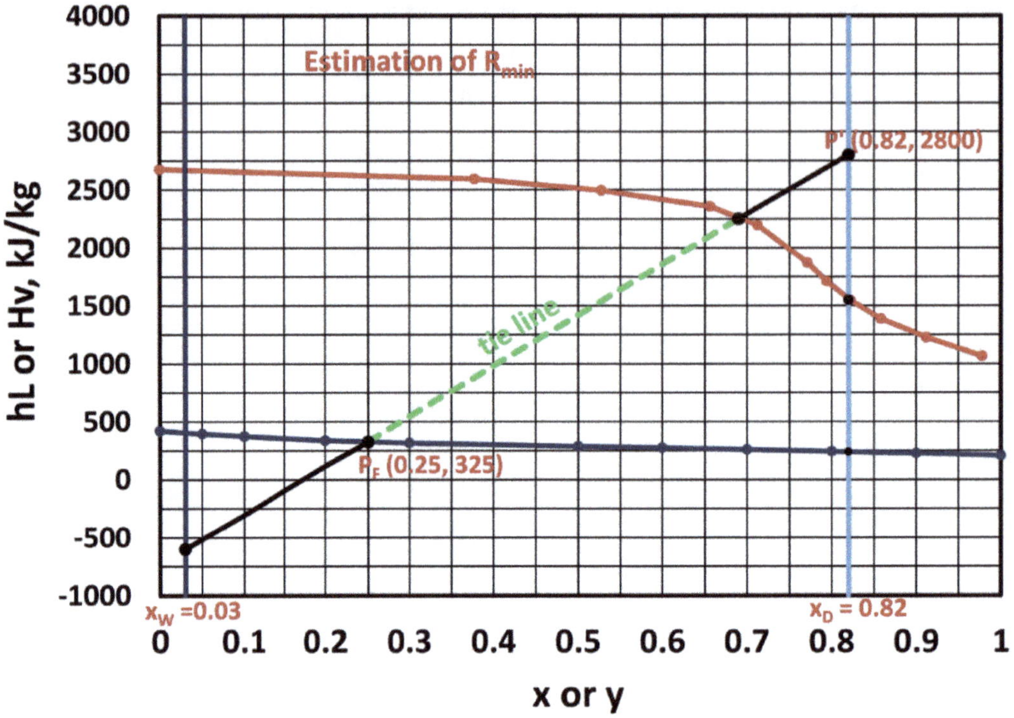

b) Number of stages required at total reflux are estimated graphically as shown below. $N \approx 3$ including partial reboiler.

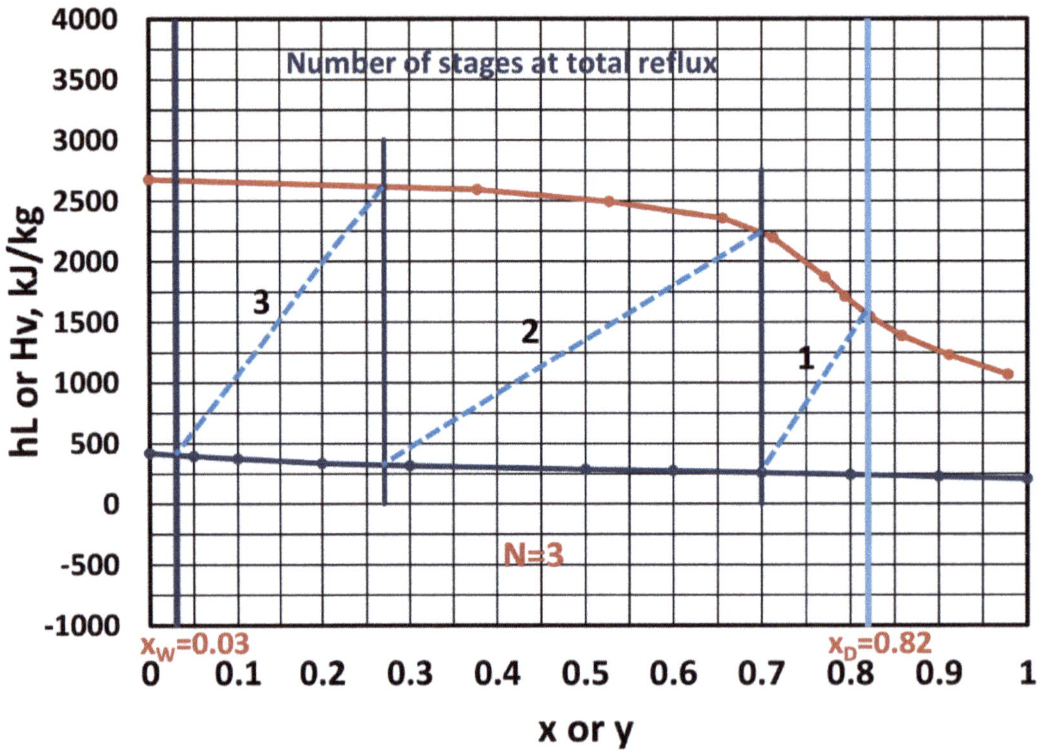

c) Condenser and reboiler heat duties at a reflux ratio of $1.30R_{min}$.

$$R = 1.30 \times 0.954 = 1.24$$

$$-\frac{\dot{Q}_C}{D} = (1 + R)\lambda_D = (1 + 1.24)(1310) = 2934.4\ kJ/kg$$

We need to calculate the distillate flow rate in order to determine the condenser heat duty. From Equation (11.41):

$$\frac{W}{D} = \frac{x_D - z_F}{z_F - x_W} = \frac{0.82 - 0.25}{0.25 - 0.03} = 2.591$$

From overall material balance:

$$F = D + W = D + 2.591D = 3.591D = 7000$$

Thus:

$$D = 1949.32\ kg/h \quad \text{and} \quad W = 5050.68\ kg/h$$

$$\dot{Q}_C = -2934.4 \times 1949.32 = -5720084.61\ \ kJ/h = -1588.91\ kW$$

The reboiler heat duty is calculated using overall energy balance, Equation (11.38):

$$Fh_F + \dot{Q}_R + \dot{Q}_C = Wh_W + Dh_D$$

$$\dot{Q}_R = Wh_W + Dh_D - Fh_F - \dot{Q}_C = (5050.68 \times 405) + (1949.32 \times 240) - (7000 \times 325)$$
$$-(-5720084.61) = 5958446.81\ kJ/h = 1655.12\ kW$$

d) The number of equilibrium stages at $R = 1.30R_{min}$ with optimum feed location.
To estimate the number of stages graphically, we need to locate the upper and lower difference points and the feed point on the Ponchon-Savarit diagram and construct stages as discussed in section 11.5.4.

$$h_{diff} = h_D - \frac{\dot{Q}_C}{D} = 240 - (-\ 5720084.61/1949.32\,) = 3174.40\ kJ/kg$$

$$\bar{h}_{diff} = h_W - \frac{\dot{Q}_R}{W} = 405 - (5958446.81/5050.68) = -774.73\ \ kJ/kg$$

The difference points and the feed point are located on the enthalpy composition diagram and stages are constructed as shown in the diagram below.

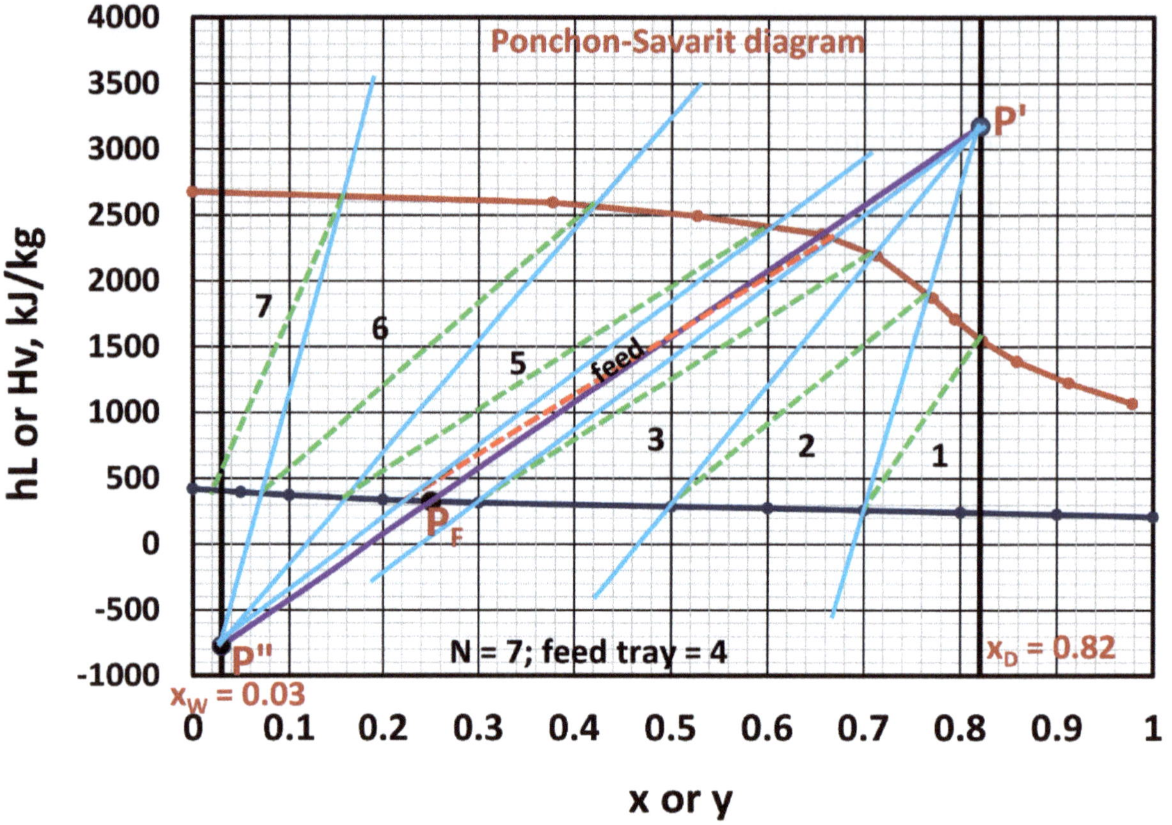

The number of stages required (including partial reboiler) is approximately 7. The optimum feed tray is 4.

Problem 2

A mixture of methanol and water is to be separated using a multistage distillation column at 1 atm. The feed is saturated liquid consisting of 50 mol% methanol. The desired distillate composition is 95 mol% methanol. The residue (bottoms product) composition is 5 mol% methanol. The reflux ratio is 2.0.
Using the Ponchon-Savarit method, estimate the number of equilibrium stages required and the location of the optimum feed tray. Assume that column uses a total condenser and a partial reboiler.
The equilibrium data and enthalpy-concentration data for mixtures of methanol and water at 1 atm pressure are given below.

Composition, mole fraction of methanol		**Enthalpy-concentration data, BTU/lb-mole**		
Liquid (x)	**Vapor (y)**	**x or y**	**Liquid, h_L**	**Vapor, H_V**
0	0	0	3240	20,720
0.02	0.134	0.05	3070	20,520
0.04	0.23	0.10	2950	20,340
0.06	0.304	0.15	2850	20,160
0.08	0.365	0.20	2760	20,000
0.10	0.418	0.30	2620	19,640
0.15	0.517	0.40	2540	19,310
0.20	0.579	0.50	2470	18,970
0.30	0.665	0.60	2410	18,650
0.40	0.729	0.70	2370	18,310

0.50	0.779	0.80	2330	17,980
0.60	0.825	0.90	2290	17,680
0.70	0.870	1.0	2250	17,390
0.80	0.915			
0.90	0.958			
0.95	0.979			
1.0	1.0			

Solution

Refer to column schematic, $y\ vs.\ x$ equilibrium curve, and enthalpy-composition diagram shown below.

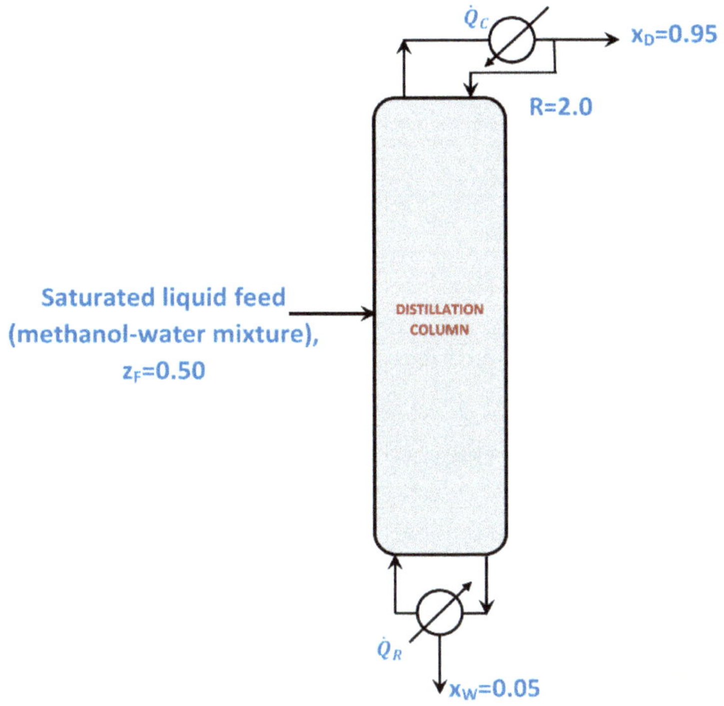

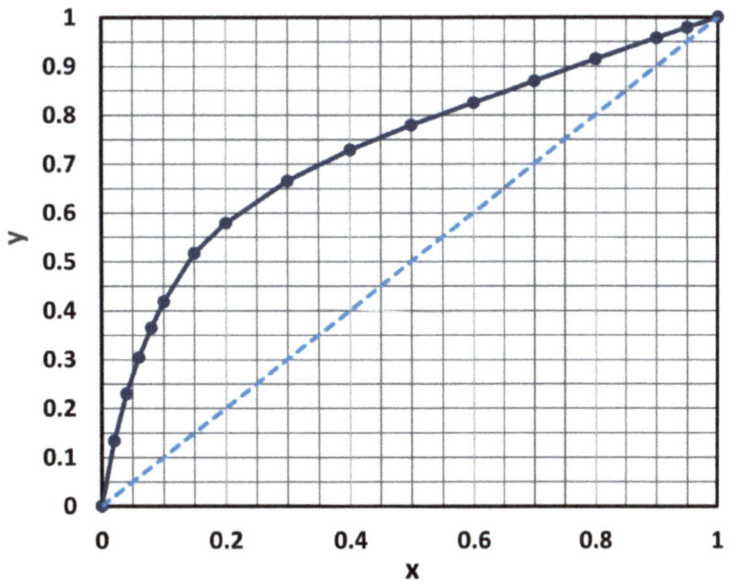

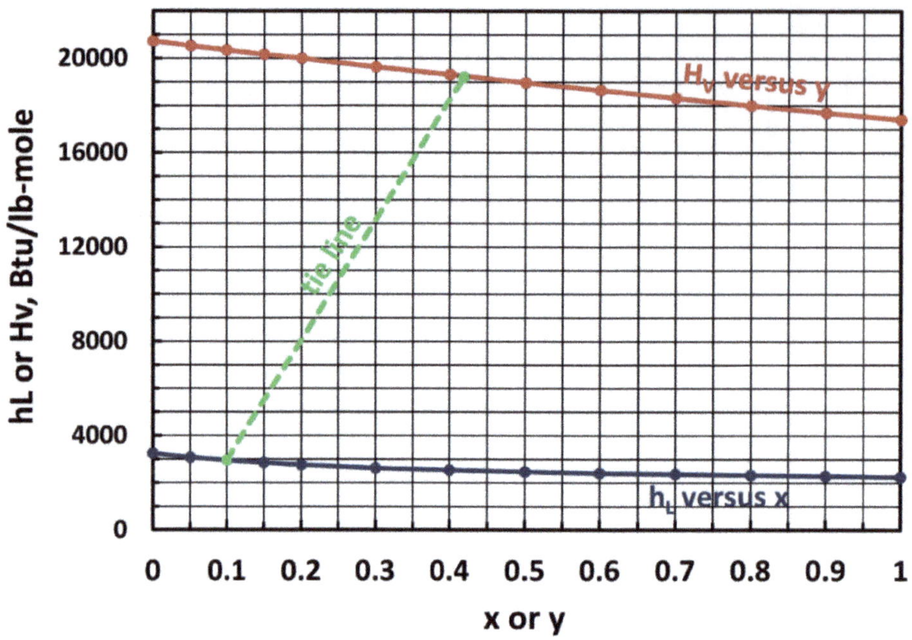

Given: Feed as saturated liquid, $R = 2.0$, $z_F = 0.50$, $x_D = 0.95$, $x_W = 0.05$. Estimate the number of equilibrium stages required and the location of the optimum feed tray. Let us first estimate the difference point enthalpies:

$$h_{diff} = (1+R)\lambda_D + h_D = (1+2)(17500 - 2250) + 2250 = 48000\ Btu/lb - mol$$

Note that λ_D and h_D are obtained from the enthalpy-composition diagram. We can calculate $\bar{h}_{diff}$ from the following relation:

$$\frac{h_{diff} - h_F}{h_F - \bar{h}_{diff}} = \frac{x_D - z_F}{z_F - x_W} = \frac{0.95 - 0.50}{0.50 - 0.05} = 1$$

$$\frac{48000 - 2500}{2500 - \bar{h}_{diff}} = 1$$

Solving for $\bar{h}_{diff}$, we get $\bar{h}_{diff} = -43{,}000\ Btu/lb - mol$

The difference points and the feed point are located on the enthalpy-composition diagram and stages are constructed as shown in the diagram below. The number of stages required (including partial reboiler) is approximately 6. The optimum feed tray is 4.

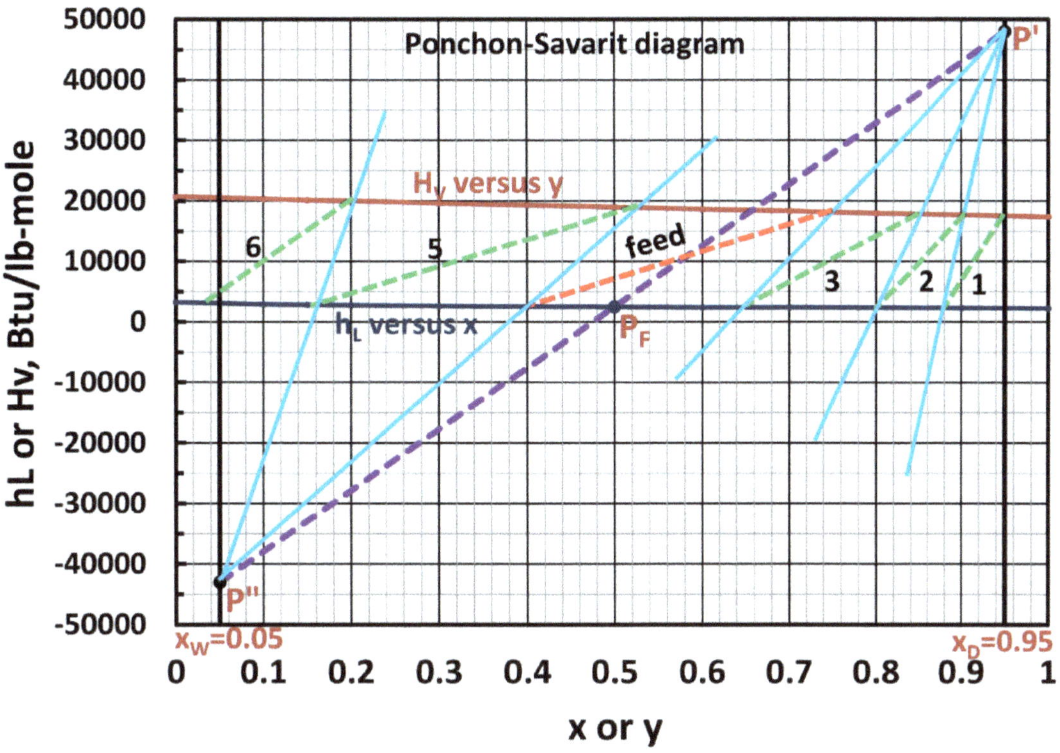

Problem 3

A multistage distillation column is to be used to separate a mixture of benzene and toluene at 1 atm pressure. The feed consists of 40 mol% benzene and is fed to the optimum tray as saturated vapor. The distillate consists of 97 mol% benzene and the bottoms product contains 4 mol% benzene. The reflux ratio is 1.95.

Using the Ponchon-Savarit method, estimate the number of equilibrium stages required and the location of the optimum feed tray. Assume that column uses a total condenser and a partial reboiler.

The equilibrium data and enthalpy-concentration data for mixtures of benzene and toluene at 1 atm pressure are given below.

Composition, mole fraction of benzene		**Enthalpy-concentration data, kJ/kmol**	
Liquid (x)	**Vapor (y)**	**Liquid, h_L**	**Vapor, H_V**
0.999732	0.999897	0	30770.34
0.856788	0.938991	427.1883	31268.12
0.726899	0.871366	877.2112	31805.1
0.608551	0.79654	1347.025	32385.59
0.500431	0.71401	1834.049	33014.21
0.401389	0.623252	2336.089	33695.93
0.310423	0.523718	2851.283	34436.08
0.226654	0.414838	3378.04	35240.36
0.149309	0.296018	3915.006	36114.83
0.077709	0.166644	4461.024	37065.99
0.011257	0.026074	5015.105	38100.74
0.000644	0.001511	5108.74	38281.83

Solution

Refer to column schematic, $y\ vs.\ x$ equilibrium curve, and enthalpy-composition diagram shown below.

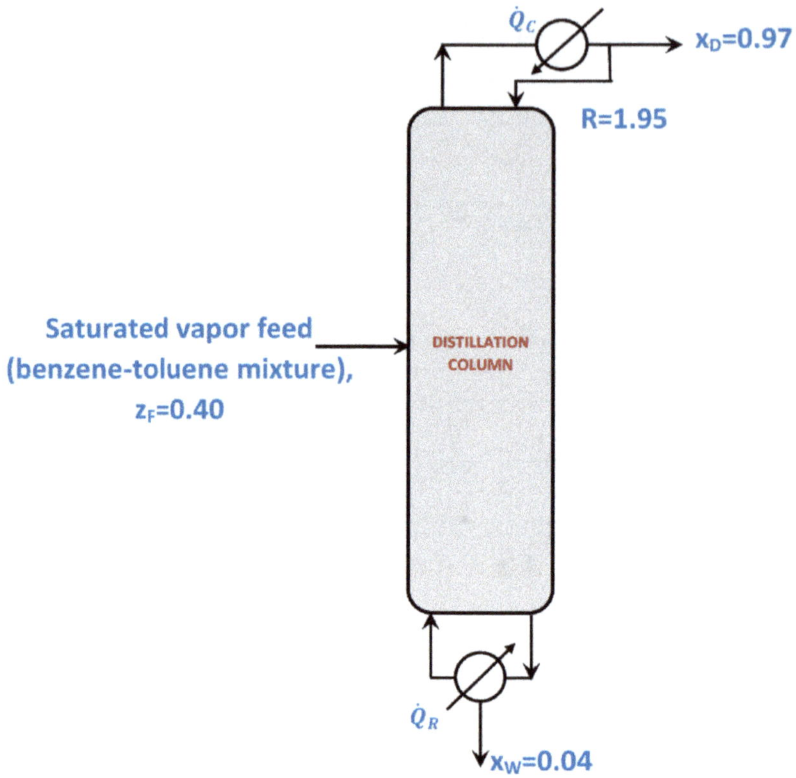

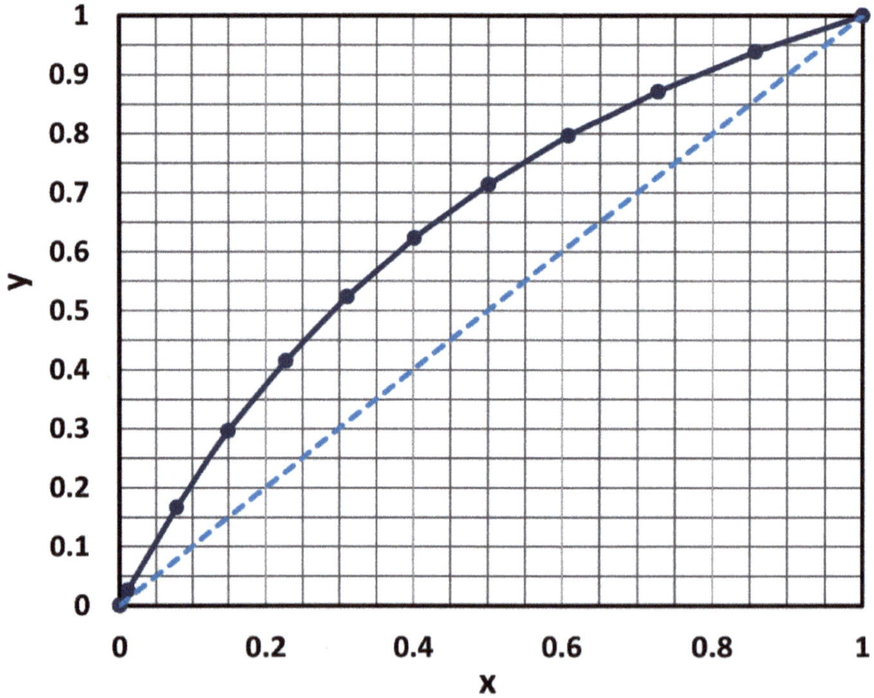

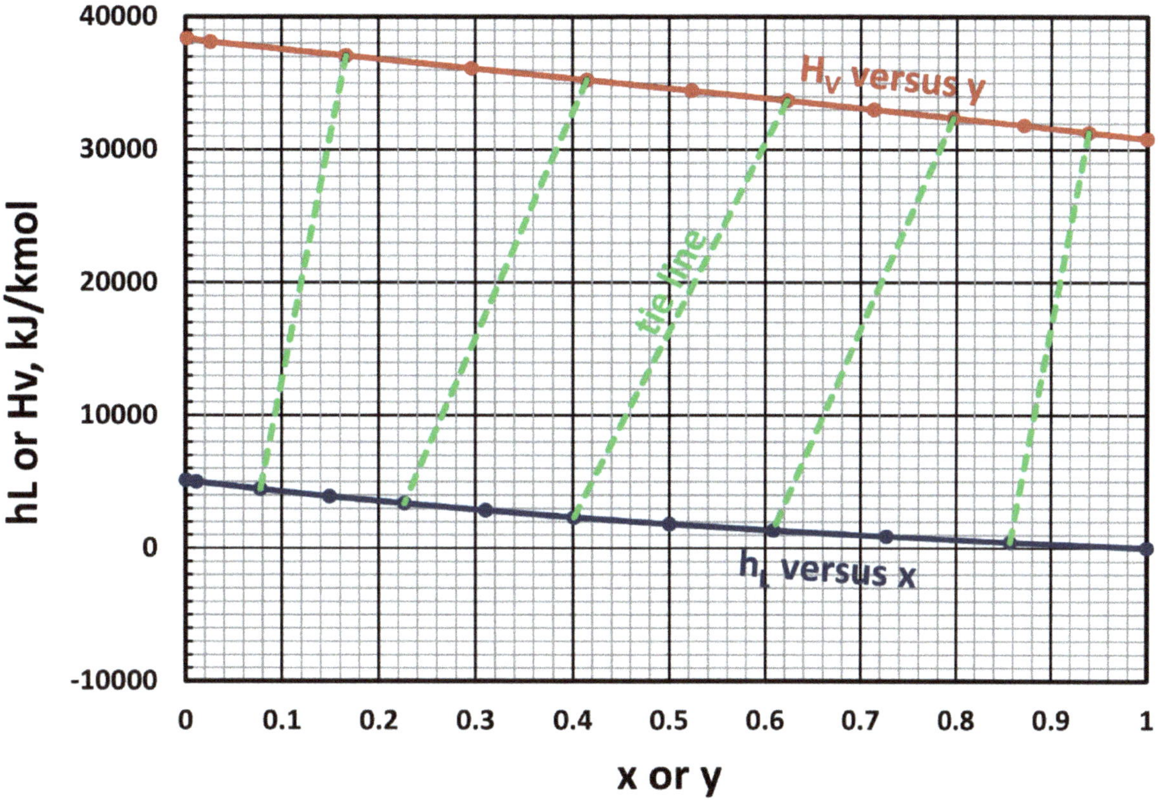

Given: Feed as saturated vapor, $R = 1.95$, $z_F = 0.40$, $x_D = 0.97$, $x_W = 0.04$. Estimate the number of equilibrium stages required and the location of the optimum feed tray. Let us first estimate the difference point enthalpies:

$$h_{diff} = (1 + R)\lambda_D + h_D = (1 + 1.95)(31000 - 80) + 80 = 91294\ kJ/kmol$$

Note that λ_D and h_D are obtained from the enthalpy-composition diagram. We can calculate $\bar{h}_{diff}$ from the following relation:

$$\frac{h_{diff} - h_F}{h_F - \bar{h}_{diff}} = \frac{x_D - z_F}{z_F - x_W} = \frac{0.97 - 0.40}{0.40 - 0.04} = 1.5833$$

$$h_{diff} - h_F = 1.5833 h_F - 1.5833 \bar{h}_{diff}$$

From the enthalpy-composition diagram, $h_F = 35200\ kJ/kmol$

Solving for $\bar{h}_{diff}$, we get $\bar{h}_{diff} = -228.54\ kJ/kmol$

The difference points and the feed point are located on the enthalpy-composition diagram. An attempt is made to construct the stages. Clearly there is no solution for this problem. As we approach the feed line, the graphical stages are becoming smaller and smaller. Obviously the given reflux ratio of 1.95 is lower than the minimum reflux ratio. Hence this is an impossible operation.

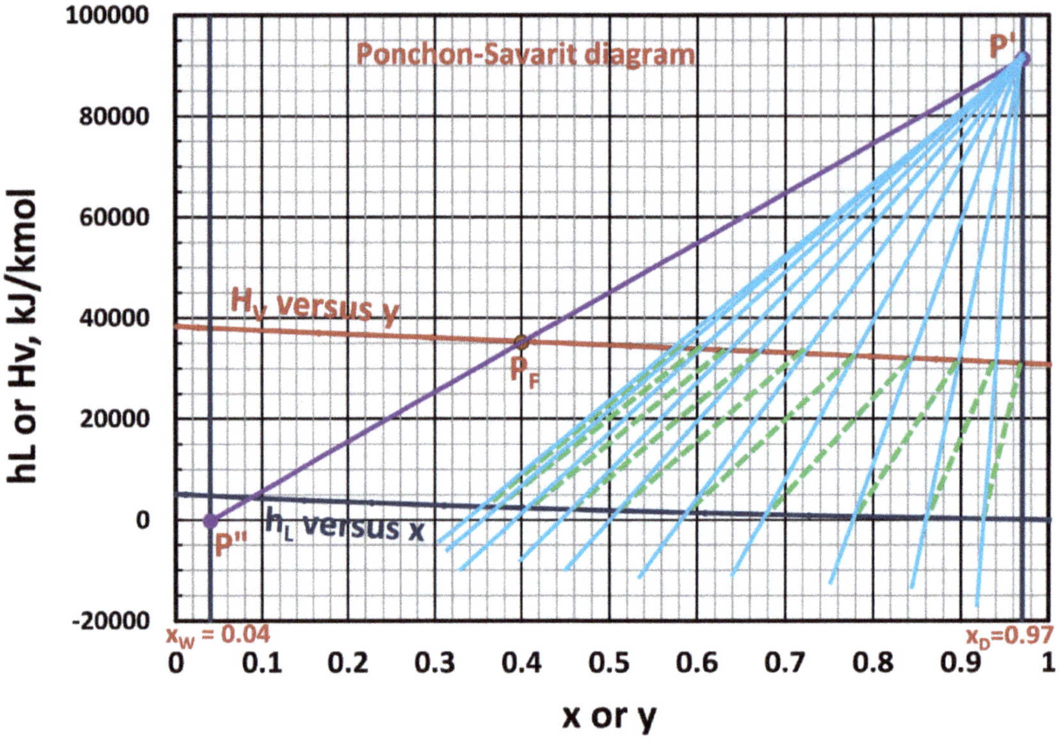

Let us re-do the problem with $R = 1.25R_{min}$

Given: Feed as saturated vapor, $z_F = 0.40$, $x_D = 0.97$, $x_W = 0.04$. Let us first estimate the minimum reflux ratio. As discussed in section 11.5.5, minimum reflux ratio is determined by extending the tie-line passing through the feed point $P_F(z_F, h_F)$ to the intersections with the two vertical lines corresponding to the product compositions ($x = x_D$ vertical line and $x = x_W$ vertical line). The extension of tie line passing through the feed point is shown in the figure given below. From the upper difference point enthalpy estimated graphically, the minimum reflux is calculated using Equation (11.42):

$$R_{min} = \left(h_{diff} - H_D\right)/\lambda_D$$

where $h_{diff} = 136{,}000\ kJ/kmol$, $H_D = 31{,}000\ kJ/kmol$, and $\lambda_D = H_D - h_D = 31000 - 80 = 30920\ kJ/kmol$. Thus,

$$R_{min} = (136000 - 31000)/30920 = 3.396$$

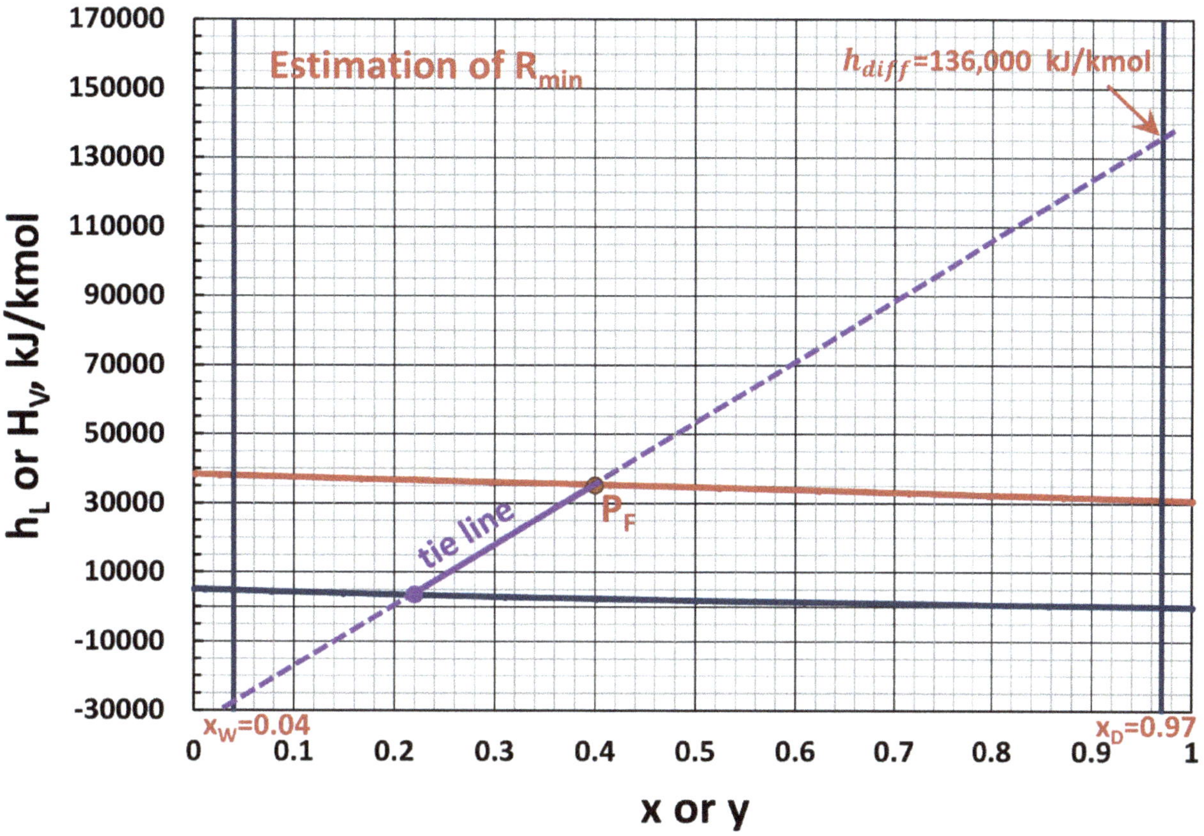

The actual reflux ratio is:

$$R = 1.25R_{min} = 1.25 \times 3.396 = 4.245$$

Given: Feed as saturated vapor, $R = 4.245$, $z_F = 0.40$, $x_D = 0.97$, $x_W = 0.04$. Estimate the number of equilibrium stages required and the location of the optimum feed tray. The difference point enthalpies are estimated as follows:

$$h_{diff} = (1 + R)\lambda_D + h_D = (1 + 4.245)(30920) + 80 = 162{,}255.4\ kJ/kmol$$

We can calculate $\bar{h}_{diff}$ from the following relation:

$$\frac{h_{diff} - h_F}{h_F - \bar{h}_{diff}} = \frac{x_D - z_F}{z_F - x_W} = \frac{0.97 - 0.40}{0.40 - 0.04} = 1.5833$$

$$h_{diff} - h_F = 1.5833h_F - 1.5833\bar{h}_{diff}$$

From the enthalpy-composition diagram, $h_F = 35200\ kJ/kmol$

Solving for $\bar{h}_{diff}$, we get $\bar{h}_{diff} = -45{,}047.2\ kJ/kmol$

The difference points and the feed point are located on the enthalpy-composition diagram and stages are constructed as shown in the diagram below. The number of stages required (including partial reboiler) is approximately 12. The optimum feed tray is 8.

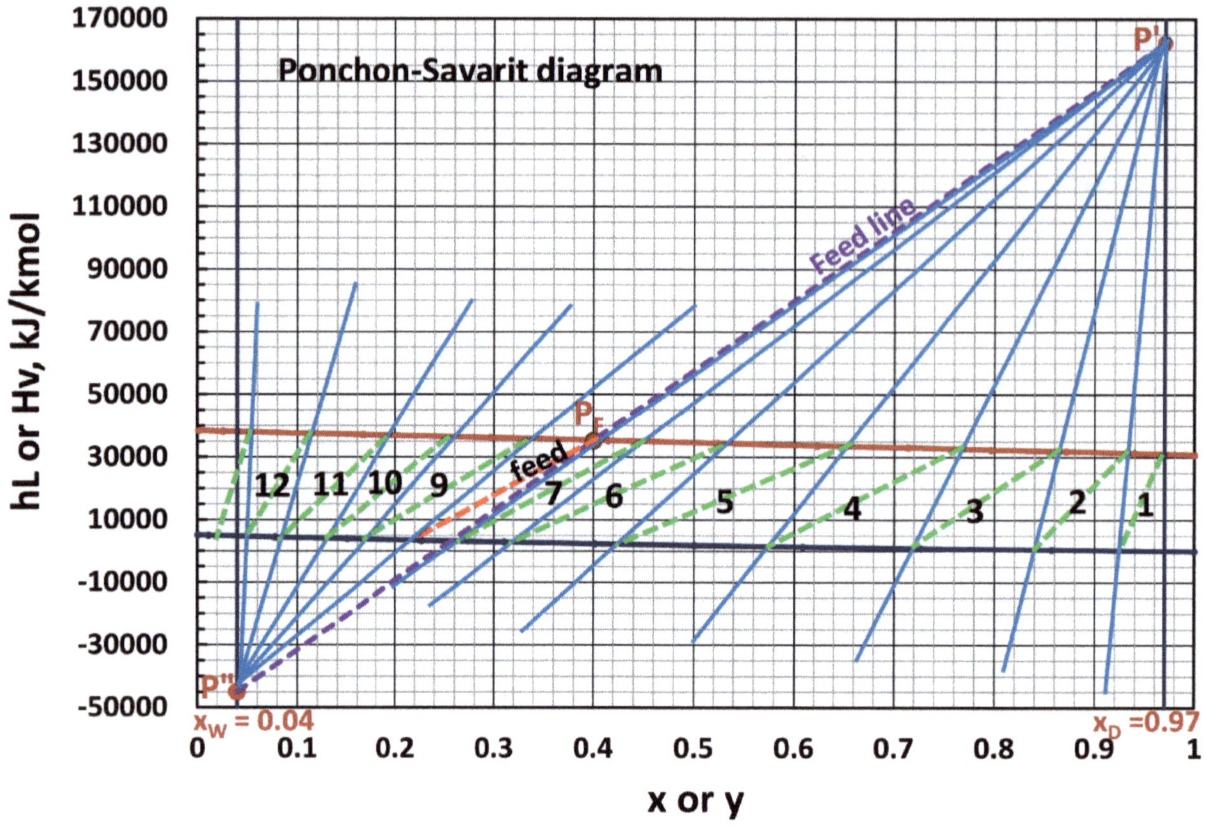

The same problem is now solved using the McCabe-Thiele approach assuming CMO. As shown in the McCabe-Thiele diagram given below, the number of stages required (including partial reboiler) is 12 and the optimum feed tray is 8. Thus the McCabe-Thiele approach gives the same results as obtained using the Ponchon-Savarit method. This is not surprising as benzene and toluene have similar values of latent heats of vaporization ($\lambda_{benzene} = 7.36\ kcal/mol$ and $\lambda_{toluene} = 7.96\ kcal/mol$ at normal boiling points) and hence, the assumption of CMO is valid.

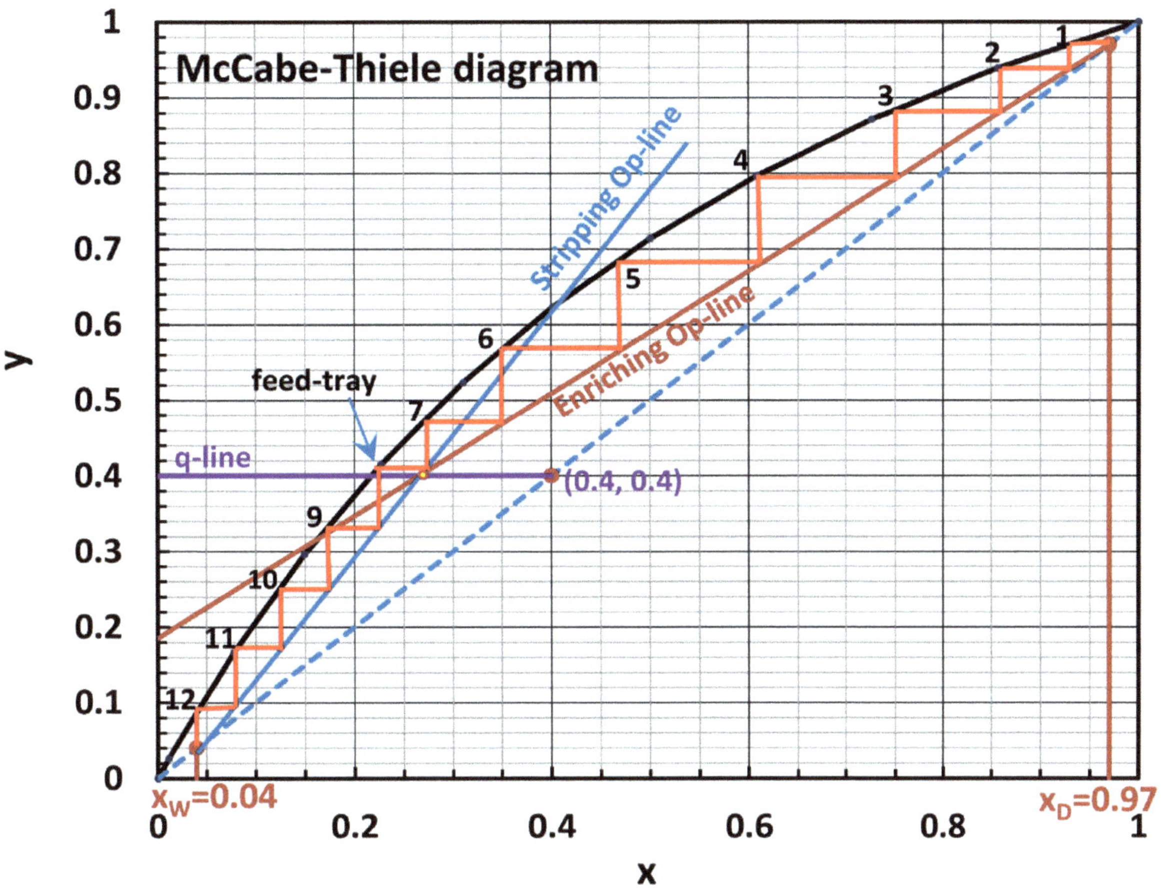
McCabe-Thiele diagram
Stripping Op-line
Enriching Op-line
feed-tray
q-line
(0.4, 0.4)
x_W=0.04
x_D=0.97
y
x
1
2
3
4
5
6
7
9
10
11
12
0
0.1
0.2
0.3
0.4
0.5
0.6
0.7
0.8
0.9
1
0
0.2
0.4
0.6
0.8
1

Chapter 12

Batch Distillation - Without and With Reflux

Problem 1

A simple batch still is used to separate ethanol from water. The still is loaded with 100 kmoles of 23 mol% ethanol mixture. The distillation is carried out at 1 atm pressure. At the end of distillation, the still pot composition is 3 mol% ethanol.
Do the following:

a) Calculate the amount of distillate collected.
b) Calculate the amount of residue left in the still pot.
c) Calculate the average composition of distillate

The equilibrium data for ethanol-water system at 1 atm pressure are given below:

Composition, mole fraction of ethanol	
Liquid (x)	**Vapor (y)**
0	0
0.019	0.17
0.0721	0.3891
0.0966	0.4375
0.1238	0.4704
0.1661	0.5089
0.2377	0.5445
0.2608	0.558
0.3273	0.5826
0.3965	0.6122
0.5079	0.6564
0.5198	0.6599
0.5732	0.6841
0.6763	0.7385
0.7472	0.7815
0.8943	0.8943
1.0	1.0

Solution

Given: $F = 100$ kmoles, $x_F = 0.23$, $x_W = 0.03$. Using the Raleigh equation (Equation (12.7))

$$ln\left(\frac{W}{F}\right) = -\int_{0.03}^{0.23} \frac{dx}{y-x} = -Area$$

The data and plot of $1/(y-x)$ versus x data are shown below. Note that ethanol-water system forms an azeotrope at high ethanol concentration. Consequently, the plot diverges at high ethanol concentration.

x	y	1/(y-x)
0	0	-
0.019	0.17	6.62252
0.0721	0.3891	3.15457
0.0966	0.4375	2.93341
0.1238	0.4704	2.88517
0.1661	0.5089	2.91715
0.2377	0.5445	3.25945
0.2608	0.558	3.36474
0.3273	0.5826	3.91696
0.3965	0.6122	4.63607
0.5079	0.6564	6.73401
0.5198	0.6599	7.13776
0.5732	0.6841	9.01713
0.6763	0.7385	16.0772
0.7472	0.7815	29.1545
0.8943	0.8943	-
1.0	1.0	-

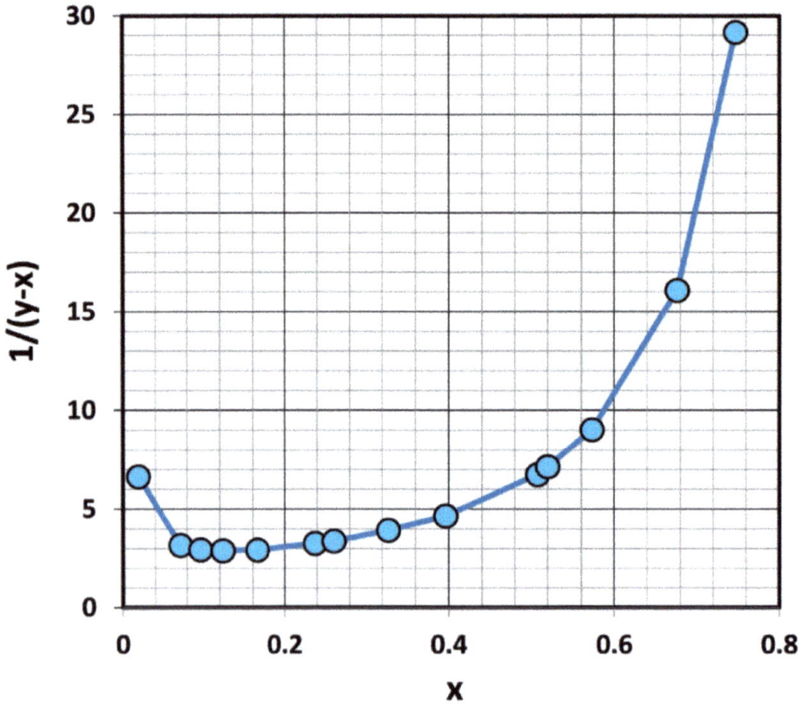

The area under the curve $1/(y-x)$ versus x curve is determined (see the figure below) using the trapezoidal rule. The total area under the curve is approximated by 5 trapezoids as shown in the figure below. The total area of 5 trapezoids is as follows:

$$Area = (0.0721 - 0.03) \times \left(\frac{3.1546 + 5.9}{2}\right) + (0.0966 - 0.0721) \times \left(\frac{3.1546 + 2.9334}{2}\right)$$
$$+ (0.1238 - 0.0966) \times \left(\frac{2.9334 + 2.88517}{2}\right)$$
$$+ (0.1661 - 0.1238) \times \left(\frac{2.88517 + 2.9172}{2}\right) + (0.23 - 0.1661) \times \left(\frac{2.9172 + 3.2}{2}\right)$$
$$= 0.6624$$

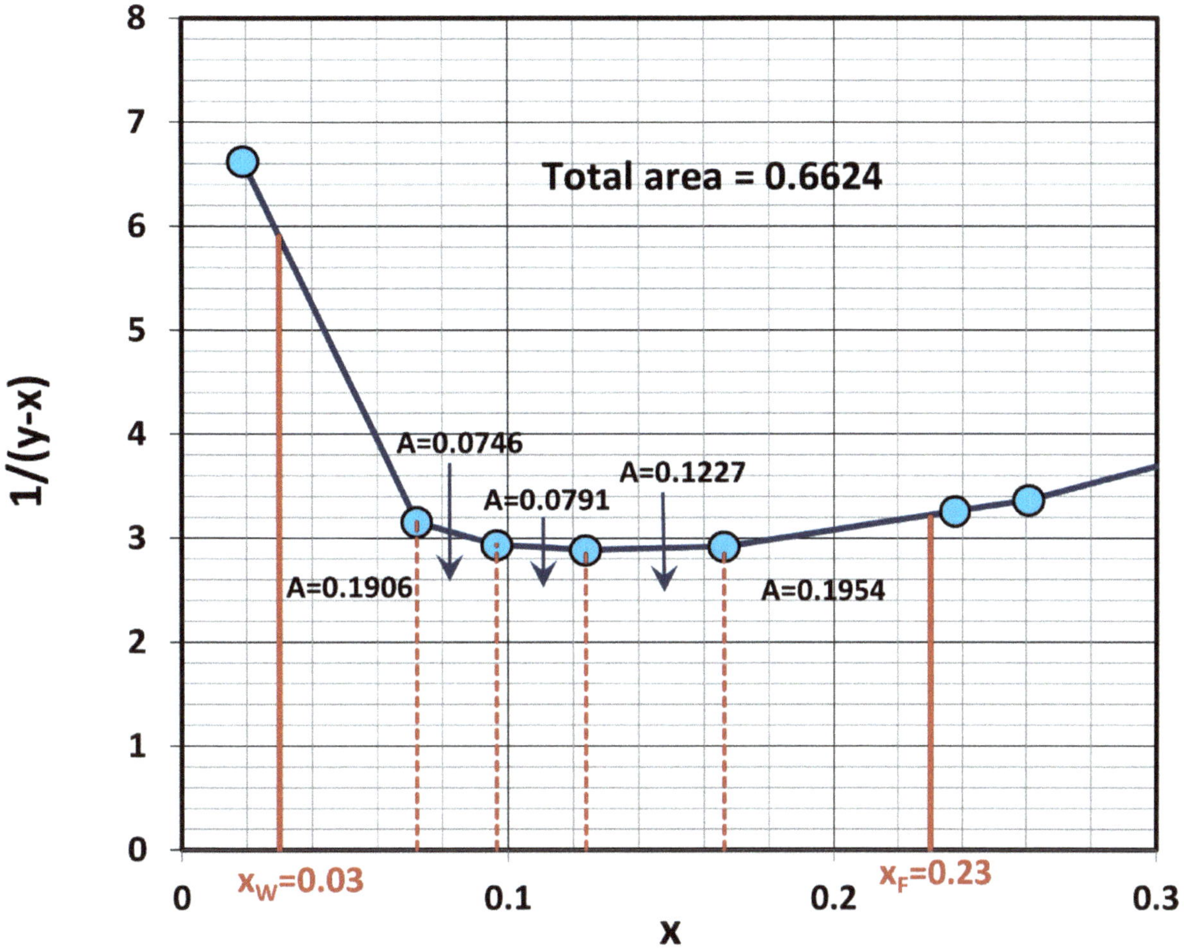

Thus:

$$ln\left(\frac{W}{F}\right) = -0.6624$$

$$\frac{W}{F} = 0.5156$$

$$W = 0.5156 \times 100 = 51.56 \text{ kmoles}$$

$$D = F - W = 100 - 51.56 = 48.44 \text{ kmoles}$$

$$x_D = \frac{x_F F - x_W W}{D} = \frac{0.23 \times 100 - 0.03 \times 51.56}{48.44} = 0.443$$

Problem 2

50 kmoles of a binary mixture of benzene and toluene containing 40 mol% benzene is loaded in a batch still. The liquid boilup rate in the still pot (-dW/dt) is maintained at 4 kmol/h. After 5 hours of distillation, the composition of liquid left inside the pot is 19 mol% benzene.

a) Calculate the amount of liquid left inside the pot after 5 hours.
b) Calculate the amount of distillate collected after 5 hours.
c) Calculate the average composition of distillate after 5 hours.

d) Estimate the relative volatility using the Rayleigh equation.

Solution

Given: $F = 50$ kmoles, $x_F = 0.40, -dW/dt = 4$ kmol/h, $x_W = 0.19$ after 5 h of distillation.

$$F = D + W$$

$$dF/dt = dD/dt + dW/dt = 0$$

$$dD/dt = -dW/dt = 4 \text{ kmol/h}$$

Upon integration and using the initial condition that $W = F = 50$ kmoles, the following result is obtained:

$$W = 50 - 4t$$

a) $W = 50 - 4t = 50 - 4 \times 5 = 30$ kmoles

b) $D = F - W = 50 - 30 = 20$ kmoles

c) $x_D = \frac{x_F F - x_W W}{D} = \frac{0.40 \times 50 - 0.19 \times 30}{20} = 0.715$

d) To estimate the relative volatility, the following Rayleigh equation, Equation (12.17), is used:

$$\frac{n_A}{n_{A0}} = \left[\frac{n_B}{n_{Bo}}\right]^{\alpha}$$

$n_{Ao} = 50 \times 0.40 = 20$ kmoles, $n_{Bo} = 50 - 20 = 30$ kmoles, $n_A = 0.19 \times 30 = 5.7$ kmoles, and $n_B = 0.81 \times 30 = 24.3$ kmoles. Apply the above Rayleigh equation:

$$\frac{5.7}{20} = \left[\frac{24.3}{30}\right]^{\alpha}$$

$$\alpha = \frac{ln\left(\frac{5.7}{20}\right)}{ln\left(\frac{24.3}{30}\right)} = 5.96$$

Comments:

The relative volatility of benzene to toluene at 1 atm is approximately 2.5. The calculated value is more than double the actual value. Clearly, the data presented in the problem is questionable. Nevertheless, the calculations and the calculation procedure are correct.

Problem 3

An equimolar mixture of benzene and toluene is fed to a batch still and distilled at 1 atm pressure. The amount of mixture fed is 75 kmoles. The liquid boilup rate in the still pot (-dW/dt) is maintained at 7.5 kmol/h. The relative volatility of benzene with respect to toluene is nearly constant and has a value of 2.40. Do the following:

a) Plot instantaneous vapor composition as a function of time.

b) Plot still pot composition as a function of time.

c) Plot average distillate composition as a function of time.

Solution

Given: $F = 75$ kmoles, $x_F = 0.50$, $-dW/dt = 7.5$ kmol/h, $\alpha = 2.40$.

Equation (12.16) gives:

$$ln\left(\frac{W}{F}\right) = \frac{1}{(\alpha - 1)} ln\left[\frac{x_W(1 - x_F)}{x_F(1 - x_W)}\right] + ln\left[\frac{1 - x_F}{1 - x_W}\right]$$

$$\left(\frac{W}{F}\right) = \left[\left(\frac{x_W}{x_F}\right)\left(\frac{1 - x_F}{1 - x_W}\right)^{\alpha}\right]^{\frac{1}{(\alpha-1)}}$$

$$\left(\frac{W}{75}\right) = \left[\left(\frac{x_W}{0.50}\right)\left(\frac{1 - 0.50}{1 - x_W}\right)^{2.4}\right]^{\frac{1}{(2.4-1)}} \quad (1)$$

Upon integration of $-dW/dt = 7.5$ using the initial condition that $W = F = 75$ kmoles, the following result is obtained:

$$W = 75 - 7.5t \quad (2)$$

From Equation (12.14):

$$y = \alpha x_w/[1 + (\alpha - 1)x_w] \quad (3)$$

$$x_{D,avg} = \frac{x_F F - x_W W}{D} \quad (4)$$

Procedure to calculate and plot various quantities indicated in parts (a) to (c) of the problem:

- Select a number of x_w values in the range of 0 to 0.50.
- Calculate W from Equation (1)
- Calculate corresponding time t from Equation (2)
- Calculate y from Equation (3)
- Calculate $x_{D,avg}$ from Equation (4)

The plots of various quantities are shown in the figure given below. All quantities decrease with the progress of distillation.

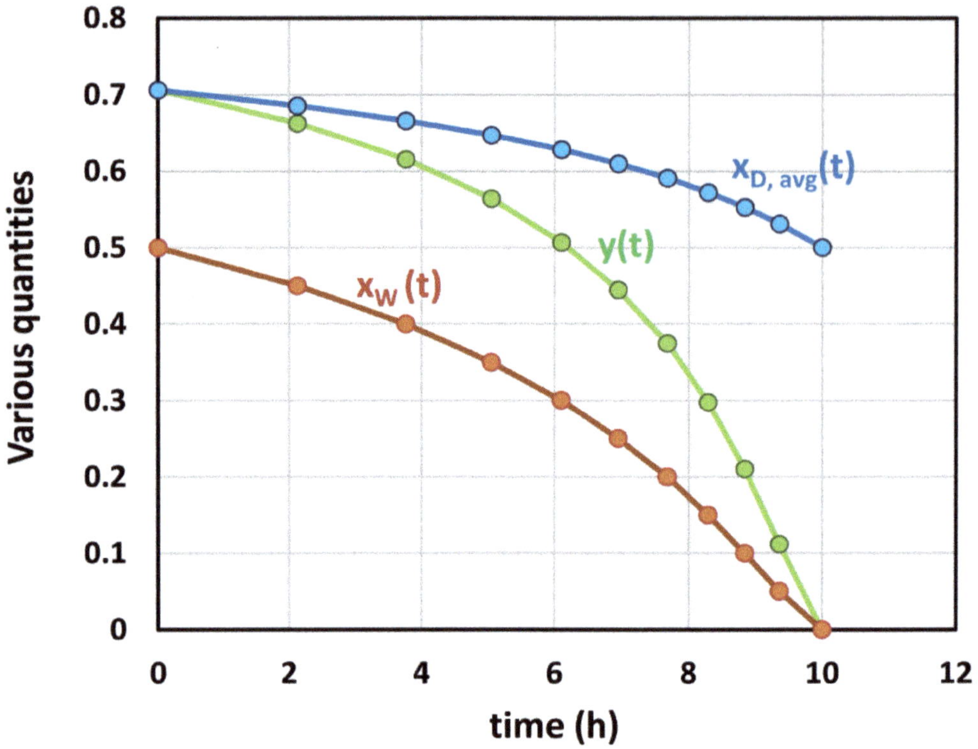

Problem 4

Batch distillation with reflux is used to separate a mixture of ethanol and water. The distillation system consists of a still pot, a column with two equilibrium stages, and a total condenser. The feed to the still pot is 40 mol% ethanol mixture. The amount of initial feed is 100 kmoles. The reflux ratio is 0.75. The system is operated at 1 atm pressure. The final residue composition in the still pot is 6 mol% ethanol.

a) Calculate the amount of distillate collected.
b) Calculate the amount of residue left in the still pot.
c) Calculate the average composition of distillate

The equilibrium data for ethanol-water system at 1 atm pressure are given in problem 1.

Solution

Refer to the schematic given below.

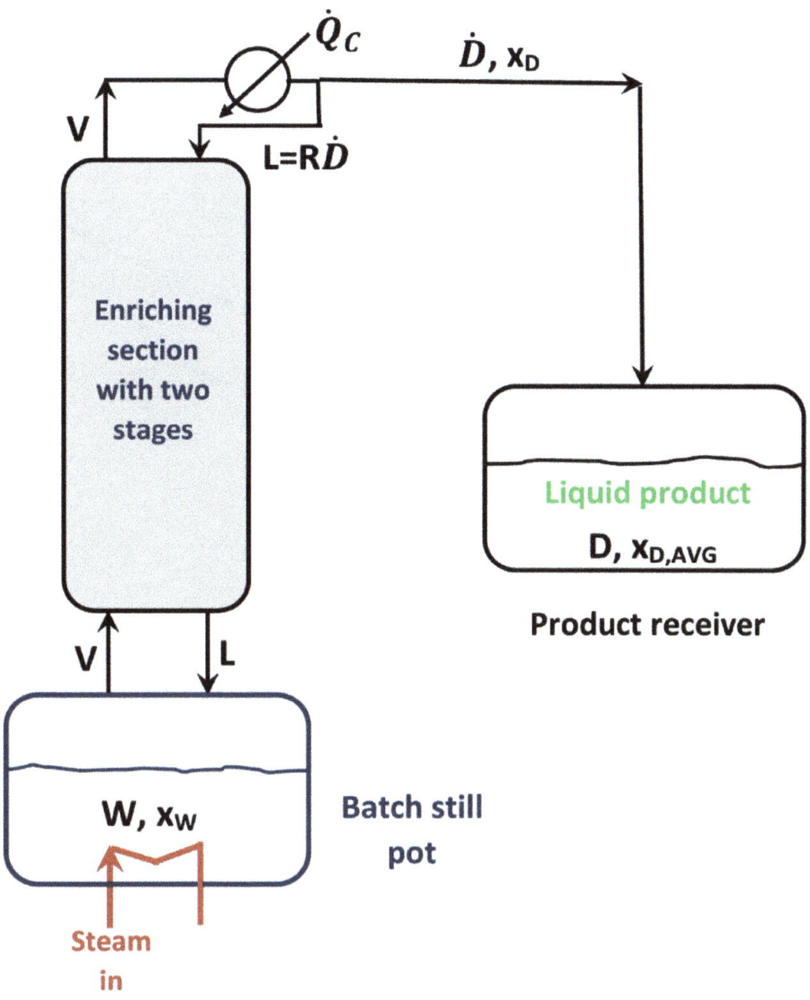

Given: $F = 100$ kmoles, $x_F = 0.40$, $R = 0.75$, $x_{W,final} = 0.06$.

As discussed in section 12.4 of the book, we need to apply the modified Rayleigh equation, Equation (12.20):

$$ln\left(\frac{W}{F}\right) = -\int_{x_{W,final}}^{x_F} \frac{dx_W}{x_D - x_W} = -Area$$

We first need to generate $1/(x_D - x_W)$ versus x_W data. We select a number of x_D values and plot the corresponding operating lines. For a fixed reflux ratio (R=0.75), all the operating lines are parallel with y-intercepts of $x_D/(1 + R)$. The family of operating lines is shown in the figure given below.

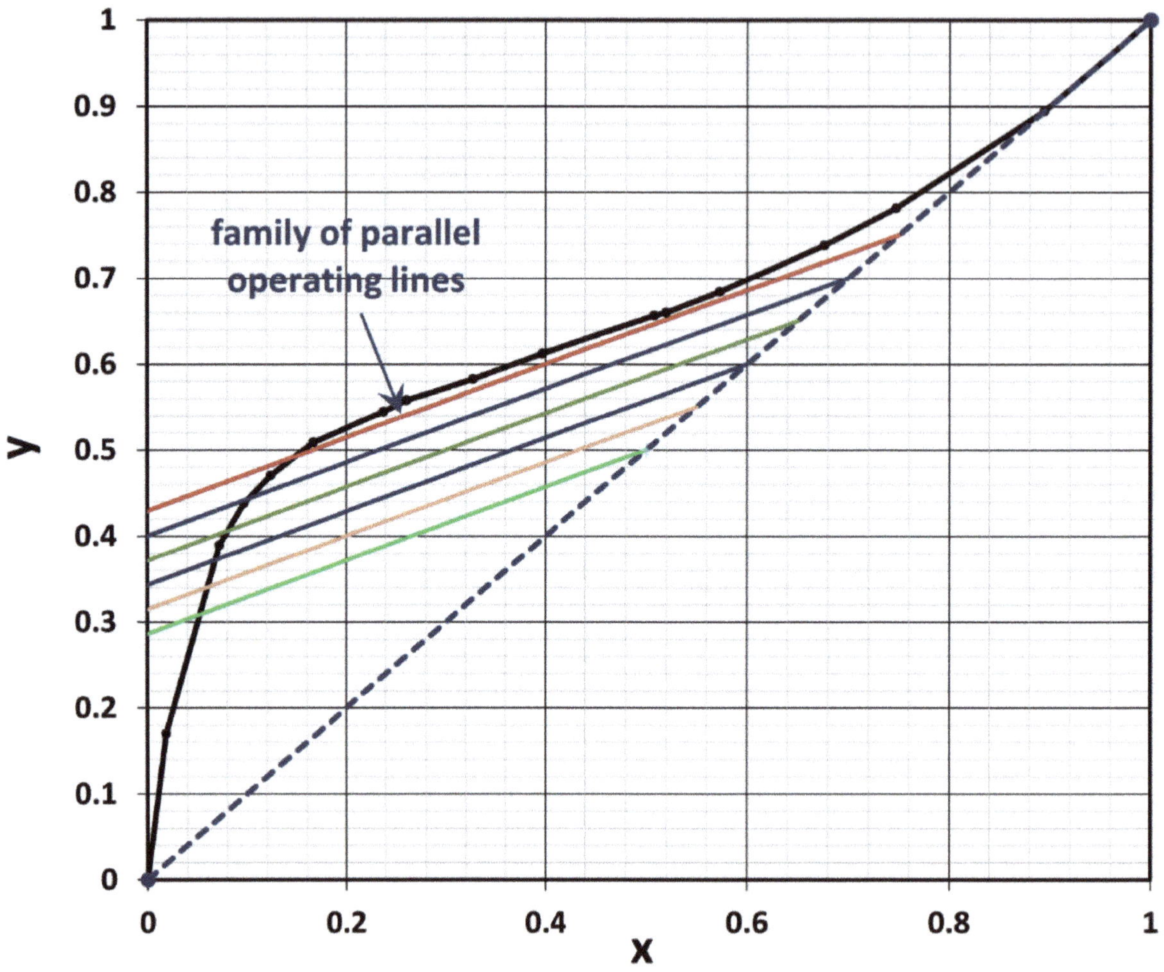

For any given operating line, the x_W value is determined by stepping off 3 stages (2 stages within the column and 1 still pot stage) as illustrated in the figure given below. The data of $1/(x_D - x_W)$ versus x_W are plotted and the area under the curve $1/(x_D - x_W)$ versus x_W is determined using the trapezoidal rule as shown in the figure given below.

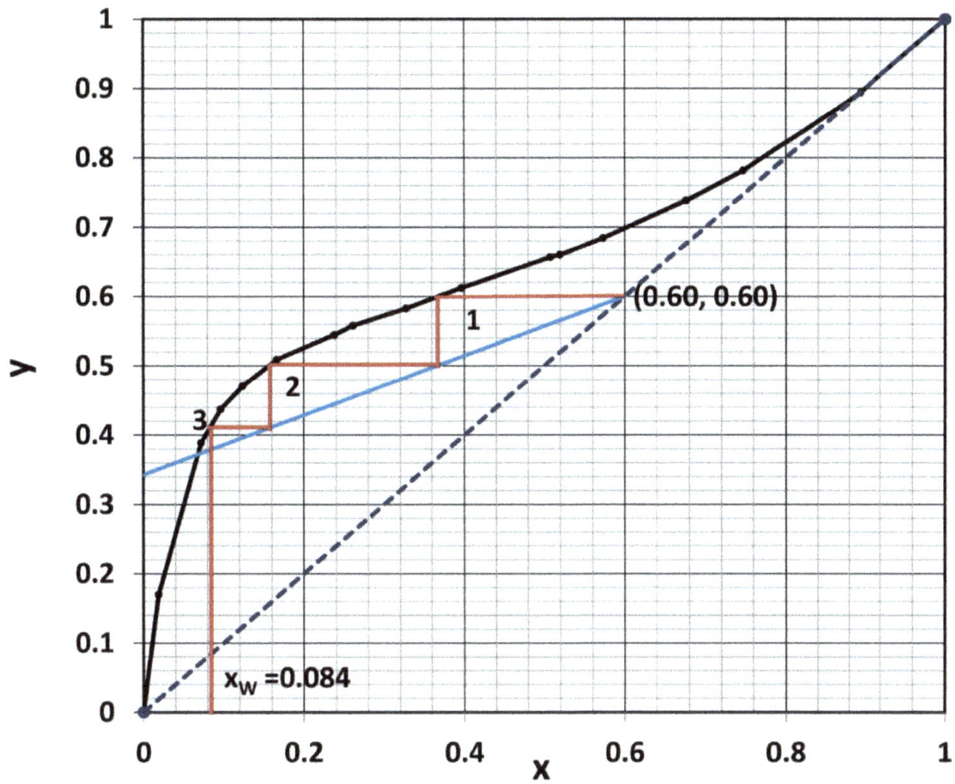
(0.60, 0.60)
1
2
3
x_W =0.084
y
x

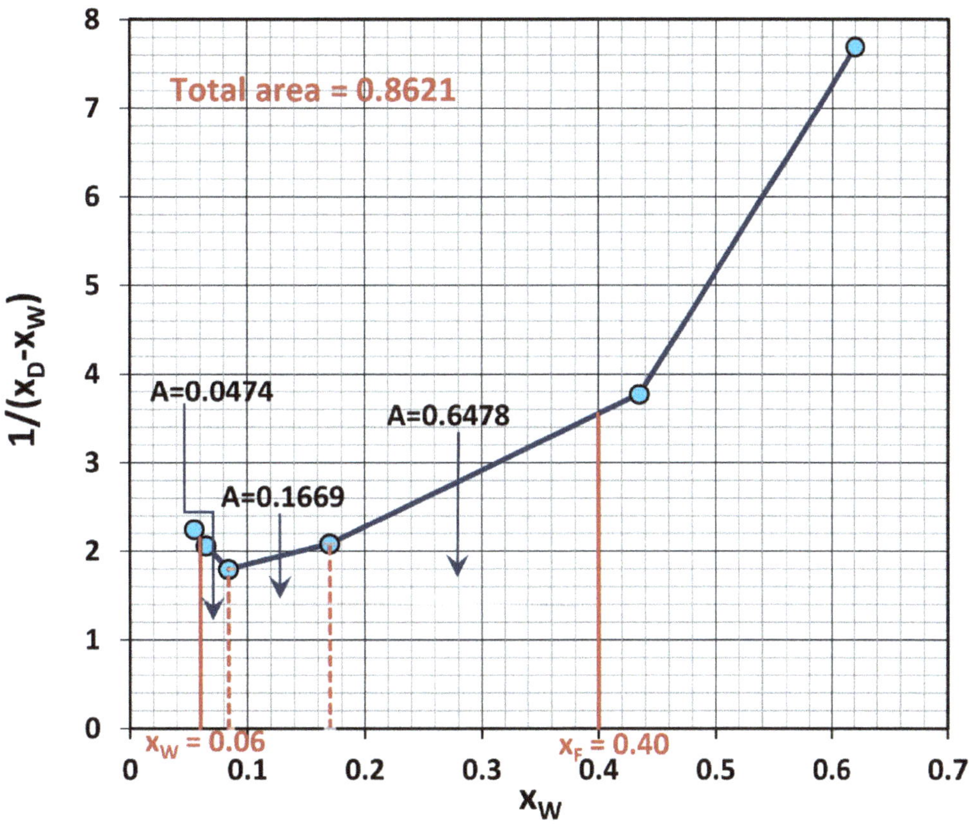
Total area = 0.8621
A=0.0474
A=0.6478
A=0.1669
x_W = 0.06
x_F = 0.40
$1/(x_D-x_W)$
x_W

The total area under the curve is approximated by 3 trapezoids as shown in the figure above. The total area of 3 trapezoids is as follows:

$$Area = (0.084 - 0.06) \times \left(\frac{1.7986 + 2.15}{2}\right) + (0.17 - 0.084) \times \left(\frac{1.7986 + 2.0833}{2}\right) + (0.40 - 0.17) \times \left(\frac{2.0833 + 3.55}{2}\right) = 0.8621$$

Thus:

$$ln\left(\frac{W}{F}\right) = -0.8621$$

$$\frac{W}{F} = 0.4223$$

$$W = 0.4223 \times 100 = 42.23 \text{ kmoles}$$

a) The amount of distillate collected is:

$$D = F - W = 100 - 42.23 = 57.77 \text{ kmoles}$$

b) The amount of residue left is 42.23 kmoles

c) The average composition of the distillate product is:

$$x_D = \frac{x_F F - x_W W}{D} = \frac{0.40 \times 100 - 0.06 \times 42.23}{57.77} = 0.6485$$

Chapter 13

Multicomponent Distillation

Problem 1

A conventional distillation column with total condenser and partial reboiler is being used to separate a 5-component mixture. The feed composition and the relative volatilities of components are given in the table below.

Component	$z_{i,F}$	$\alpha_{i,HK}$
1(LNK)	0.07	2.5
2(LK)	0.39	2.0
3(HK)	0.41	1.0
4(HNK)	0.07	0.90
5(HNK)	0.06	0.70

The fractional recovery of LK component in the distillate is 98%. The fractional recovery of HK component in the bottoms is 97%. Calculate the minimum number of equilibrium stages required? Specify the fractional recoveries of all components in the distillate and bottoms products at total reflux?

Solution

The figure below shows schematically the column operating at total reflux.

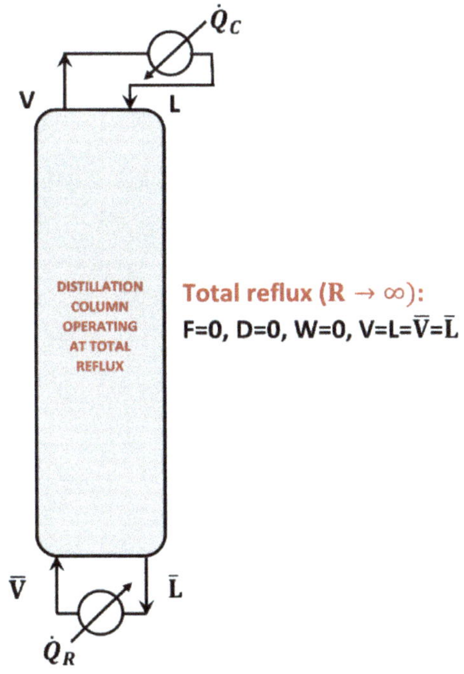

We apply Fenske equation, Equation (13.6), to calculate N_{min}

$$N_{min} = \frac{ln[\{FR_{LK}/(1-FR_{LK})\}_D/\{(1-FR_{HK})/FR_{HK}\}_W]}{ln(\alpha_{LK,HK})}$$

$$FR_{LK,D} = 0.98$$

$$FR_{HK,W} = 0.97$$

$$\alpha_{LK,HK} = 2.0$$

Thus,

$$N_{min} = \frac{ln[(0.98/(1-0.98))/((1-0.97)/0.97)]}{ln(2.0)} = 10.63$$

Using Equation (13.7), we can calculate the fractional recoveries of non-key components

$$FR_{NK,D} = \frac{\alpha_{NK,HK}^{N_{min}}}{\alpha_{NK,HK}^{N_{min}} + \{FR_{HK}/(1-FR_{HK})\}_W}$$

$$FR_{1,D} = \frac{\alpha_{1,HK}^{N_{min}}}{\alpha_{1,HK}^{N_{min}} + \{0.97/(1-0.97)\}} = \frac{(2.5)^{10.63}}{(2.5)^{10.63} + \{0.97/(1-0.97)\}} = 0.9981$$

$$FR_{4,D} = \frac{\alpha_{4,HK}^{N_{min}}}{\alpha_{4,HK}^{N_{min}} + \{0.97/(1-0.97)\}} = \frac{(0.90)^{10.63}}{(0.90)^{10.63} + \{0.97/(1-0.97)\}} \approx 0.01$$

$$FR_{5,D} = \frac{\alpha_{5,HK}^{N_{min}}}{\alpha_{5,HK}^{N_{min}} + \{0.97/(1-0.97)\}} = \frac{(0.70)^{10.63}}{(0.70)^{10.63} + \{0.97/(1-0.97)\}} \approx 0.0007$$

The fractional recoveries of different components in the bottoms product is simply:

$$FR_{i,W} = 1 - FR_{i,D}$$

The following table summarizes the fractional recoveries of all components in the distillate and bottoms products at total reflux.

Component	$FR_{i,D}$	$FR_{i,W}$
1(LNK)	0.9981	0.0019
2(LK)	0.98	0.02
3(HK)	0.03	0.97
4(HNK)	0.01	0.99
5(HNK)	0.0007	0.9993

Problem 2

A saturated liquid feed ($q = 1.0$) is fed at a rate of 100 kmol/h to a conventional distillation column with total condenser and partial reboiler. The composition of the feed is 3 mol% n-pentane (C5), 41 mol% n-hexane (C6), 48 mol% n-heptane (C7), and 8 mol% n-octane (C8). The feed is distilled so that 95% of the n-hexane is recovered in the distillate. The amount of n-heptane recovered in the bottom product is 98%. The relative volatilities can be assumed to be constant ($\alpha_{C5,C7} = 6.5,\ \alpha_{C6,C7} = 2.5,\ \alpha_{C7,C7} = 1.0, \alpha_{C8,C7} = 0.40$).

a) Calculate the number of equilibrium stages required at total reflux and the fractional recoveries of C5 in distillate and C8 in bottoms product.
b) Calculate the minimum reflux ratio using the Underwood equations
c) Calculate the total number of equilibrium stages at $R = 1.4R_{min}$, and determine the optimum feed tray location.

Solution

Refer to the schematic diagram given below.

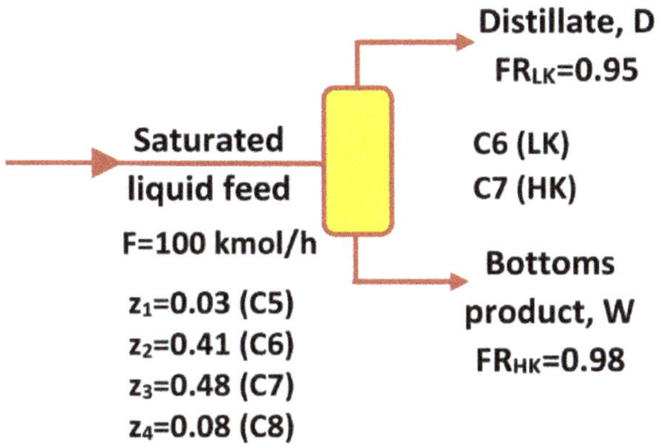

As the recoveries of n-hexane (C6) and n-heptane (C7) are specified, they are selected as the key components: more volatile n-hexane (C6) as light key (LK) and less volatile n-heptane (C7) as heavy key (HK). The information given is summarized in the table below.

Component	$z_{i,F}$	$\alpha_{i,HK}$	$FR_{i,D}$	$FR_{i,W}$
Species 1-C5(LNK)	0.03	6.5	not given	not given
Species 2- C6(LK)	0.41	2.5	0.95	0.05
Species 3 – C7 (HK)	0.48	1.0	0.02	0.98
Species 4–C8(HNK)	0.08	0.40	not given	not given

a) The number of equilibrium stages required at total reflux and fractional recoveries of C5 in distillate and C8 in bottoms product:

We apply Fenske equation, Equation (13.6), to calculate N_{min}

$$N_{min} = \frac{ln[\{FR_{LK}/(1-FR_{LK})\}_D/\{(1-FR_{HK})/FR_{HK}\}_W]}{ln(\alpha_{LK,HK})}$$

$$FR_{LK,D} = 0.95$$

$$FR_{HK,W} = 0.98$$

$$\alpha_{LK,HK} = 2.5$$

Thus,

$$N_{min} = \frac{ln[(0.95/(1-0.95))/((1-0.98)/0.98)]}{ln(2.5)} = 7.461$$

Using the other form of the Fenske equation, Equation (13.7), we can calculate the fractional recoveries of C5 (LNK) in distillate and C8 (HNK) in bottoms product as:

$$FR_{NK,D} = \frac{\alpha_{NK,HK}^{N_{min}}}{\alpha_{NK,HK}^{N_{min}} + \{FR_{HK}/(1 - FR_{HK})\}_W}$$

$$FR_{LNK,D} = \frac{\alpha_{LNK,HK}^{N_{min}}}{\alpha_{LNK,HK}^{N_{min}} + \{FR_{HK}/(1 - FR_{HK})\}_W} = \frac{(6.5)^{7.461}}{(6.5)^{7.461} + (0.98/(1 - 0.98))} = 0.99996 \approx 1.0$$

$$FR_{HNK,D} = \frac{\alpha_{HNK,HK}^{N_{min}}}{\alpha_{HNK,HK}^{N_{min}} + \{FR_{HK}/(1 - FR_{HK})\}_W} = \frac{(0.4)^{7.461}}{(0.4)^{7.461} + (0.98/(1 - 0.98))} = 2.192 \times 10^{-5} \approx 0$$

$$FR_{HNK,W} = 1 - FR_{HNK,D} = 1 - 2.192 \times 10^{-5} = 0.99998 \approx 1.0$$

b) Minimum reflux ratio using the Underwood equations:

The First Underwood equation, Equation (13.32), can be written as:

$$F(1 - q) = \sum \frac{\alpha_{i,HK}(z_i F)}{\alpha_{i,HK} - \varphi}$$

The only unknown in this equation is φ. Upon expansion of the summation term, we can write:

$$150(1 - 1) = \frac{(0.03 \times 100)(6.5)}{6.5 - \varphi} + \frac{(0.41 \times 100)(2.5)}{2.5 - \varphi} + \frac{(0.48 \times 100)(1)}{1 - \varphi} + \frac{(0.08 \times 100)(0.4)}{0.4 - \varphi}$$

$$0 = \frac{19.5}{6.5 - \varphi} + \frac{102.5}{2.5 - \varphi} + \frac{48}{1 - \varphi} + \frac{3.2}{0.4 - \varphi}$$

As discussed in section 13.3.3 (see Equation (13.35)), the solution of this equation lies within the range: $1 < \varphi < 2.5$. The solution can be obtained using the Newton-Raphson method. Another approach to solve this equation is to select several values of φ within the specified range, calculate the right hand side of the equation and plot $y = 0$ and $y = RHS$ on a y vs φ plot where RHS is the right hand side of the above equation. The intersection of the plots $y = 0$ and $y = RHS$ gives the solution. The solution is shown in the figure given below. We find $\varphi = 1.4757$ satisfies the above equation.

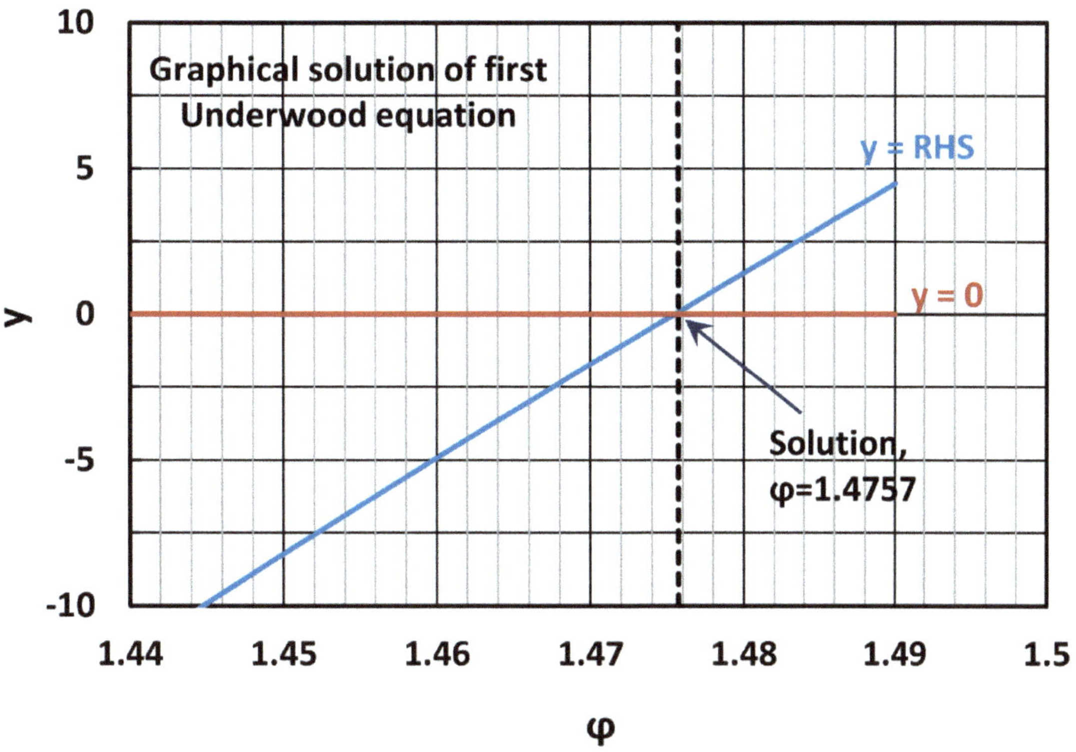

Now we use the Second Underwood equation, Equation (13.33), to determine the minimum reflux ratio R_{min}:

$$D(R_{min}+1)=\sum\frac{\alpha_{iHK}(x_{i,D}D)}{[\alpha_{iHK}-\varphi]}$$

No trial and error is needed to solve this equation for R_{min}. Before we solve this equation, we need to calculate the $(x_{i,D}D)$ terms and the total distillate flow rate D. The $(x_{i,D}D)$ terms are given as:

$$x_{i,D}D=(z_iF)(FR_{i,D})$$

Thus,

$$x_{1,D}D=(z_1F)(FR_{1,D})=(0.03\times 100)(0.99996)=2.99988\ kmol/h$$

$$x_{2,D}D=(z_2F)(FR_{2,D})=(0.41\times 100)(0.95)=38.95\ kmol/h$$

$$x_{3,D}D=(z_3F)(FR_{3,D})=(0.48\times 100)(0.02)=0.96\ kmol/h$$

$$x_{4,D}D=(z_4F)(FR_{4,D})=(0.08\times 100)(2.2\times 10^{-5})=1.76\times 10^{-4}\ kmol/h$$

$$D=2.99988+38.95+0.96+0.000176=42.91\ kmol/h$$

Now the Second Underwood equation can be solved for R_{min} as follows:

$$42.91\times(R_{min}+1)=\frac{6.5\times 2.99988}{6.5-1.4757}+\frac{2.5\times 38.95}{2.5-1.4757}+\frac{1\times 0.96}{1-1.4757}+\frac{0.40\times 0.000176}{0.40-1.4757}=96.9278$$

$$R_{min}=1.25886\approx 1.26$$

c) The total number of equilibrium stages at $R=1.4R_{min}$, and the optimum feed tray location

We can use the Gilliland approach to estimate the number of equilibrium stages at given reflux ratio.

$$R = 1.4 \times 1.25886 = 1.7624$$

$$X_G = \frac{R - R_{min}}{R+1} = \frac{1.7624 - 1.25886}{1.7624+1} = 0.18228$$

Using the Gilliland-Molokanov equation, Equation (13.41), we get

$$Y_G = 1 - exp\left[\left(\frac{X_G - 1}{\sqrt{X_G}}\right)\left(\frac{1+54.4X_G}{11+117.2X_G}\right)\right] = 1 - exp[(-1.9153)(0.3373)] = 0.4759$$

$$Y_G = \frac{N - N_{min}}{N+1} = 0.476$$

$$N = 15.147 \approx 15 \text{ stages}$$

Note that N includes partial reboiler.

The optimum feed tray location can be determined using the following Kirkbride equation:

$$\frac{n_F - 1}{N - n_F} = \left[\left(\frac{z_{HK}}{z_{LK}}\right)\left(\frac{x_{LK,W}}{x_{HK,D}}\right)^2 \frac{W}{D}\right]^{0.206}$$

where n_F is the feed stage number counting from top of the column and N is the total number of stages (including partial reboiler). $N = 15$, $z_{HK} = 0.48$, $z_{LK} = 0.41$, $D = 42.91$, and $W = F - D = 100 - 42.91 = 57.09\ kmol/h$. We need to calculate $x_{LK,W}$ and $x_{HK,D}$ as follows:

$$x_{LK,W}W = (1 - FR_{LK,D})(z_{LK}F) = (1 - 0.95)(0.41 \times 100) = 2.05$$

$$x_{LK,W} = 2.05/57.09 = 0.0359$$

$$x_{HK,D}D = (1 - FR_{HK,W})(z_{LK}F) = (1 - 0.98)(0.48 \times 100) = 0.96$$

$$x_{HK,D} = 0.96/42.91 = 0.02237$$

The Kirkbride equation can now be solved for n_F:

$$\frac{n_F - 1}{15.147 - n_F} = \left[\left(\frac{0.48}{0.41}\right)\left(\frac{0.0359}{0.02237}\right)^2 \frac{57.09}{42.91}\right]^{0.206} = 1.3313$$

$$n_F = 9.08 \approx 9$$

Thus the optimum feed tray is the 9th tray (counting from top of the column). There are 8 trays above the feed tray and 5 trays below the feed tray (excluding the partial reboiler). The partial boiler is the 15th equilibrium stage.

Problem 3

Using 4 distillation columns, show schematically the flow diagrams of 14 possible simple sequences to separate a 5-component mixture into pure components.

Solution

The flow diagrams of 14 possible simple sequences to separate a 5-component mixture into pure components are shown below.

Sequence 1

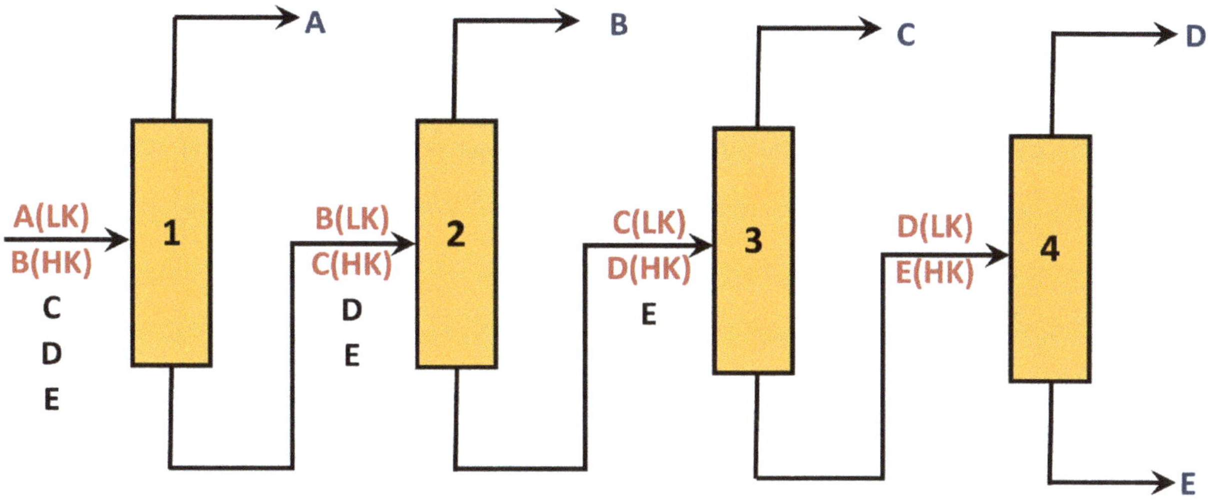

Sequence 2

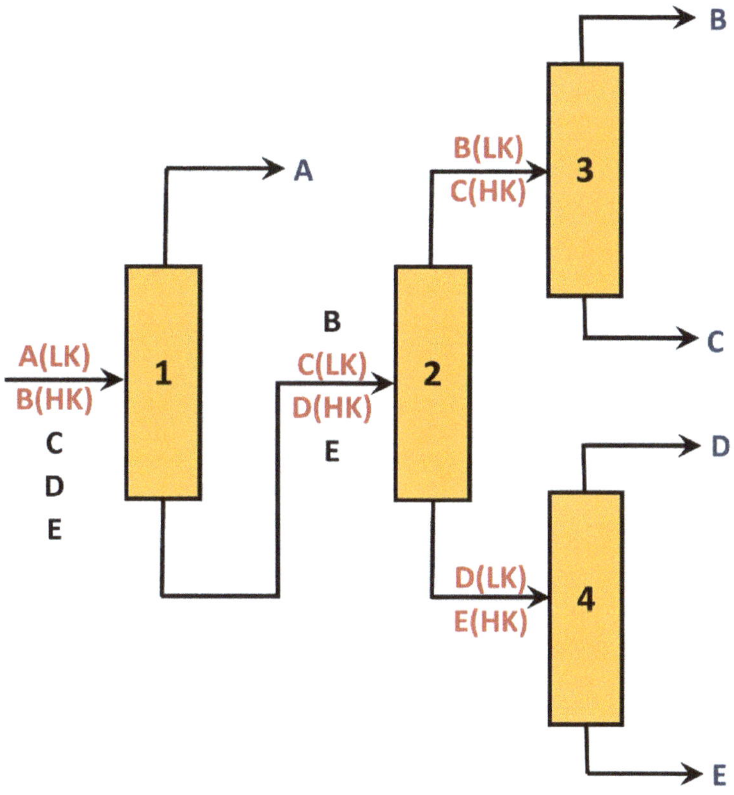

Sequence 3

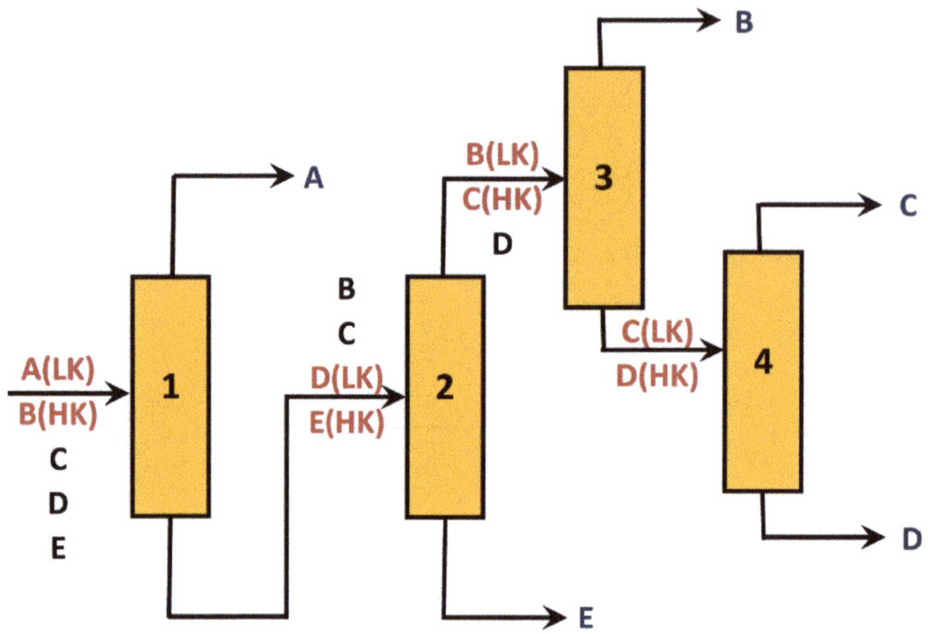

Sequence 4

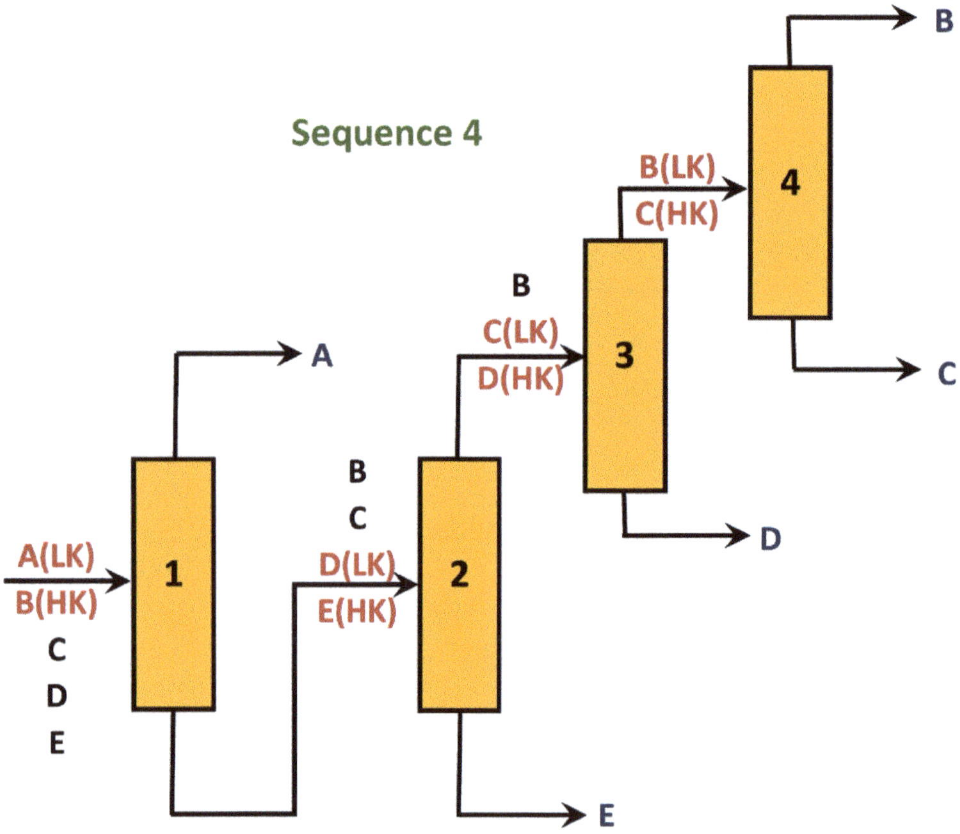

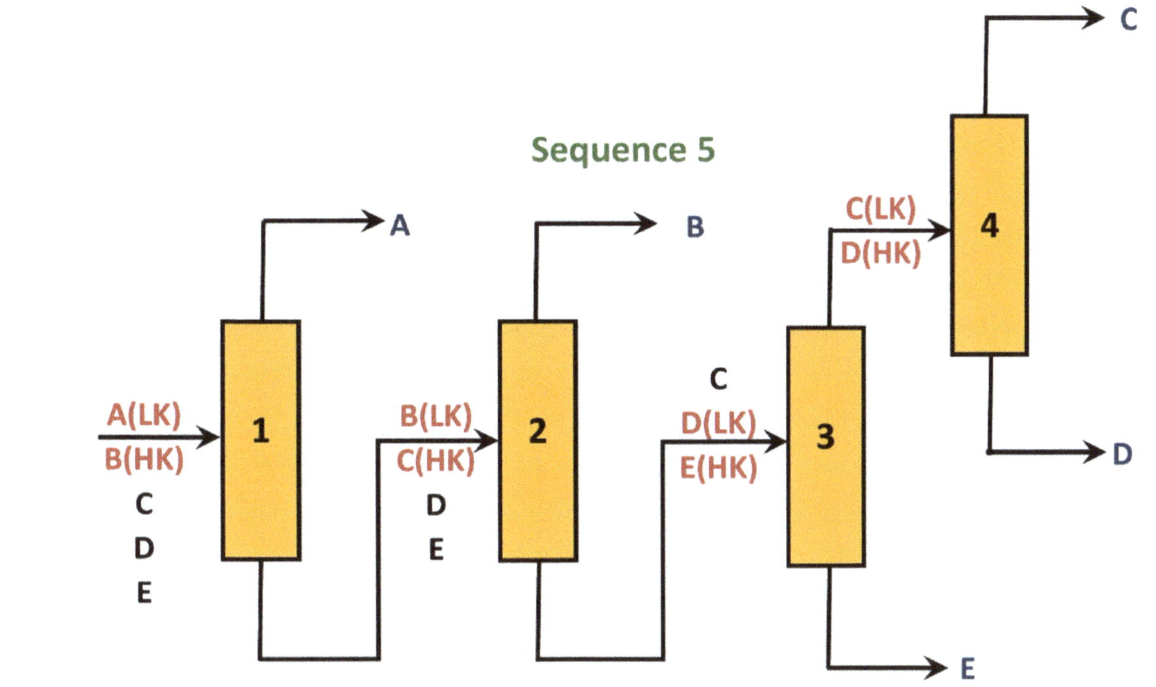
Sequence 5
A(LK)
B(HK)
C
D
E
1
A
B(LK)
C(HK)
D
E
2
B
C
D(LK)
E(HK)
3
E
C(LK)
D(HK)
4
C
D

Sequence 6

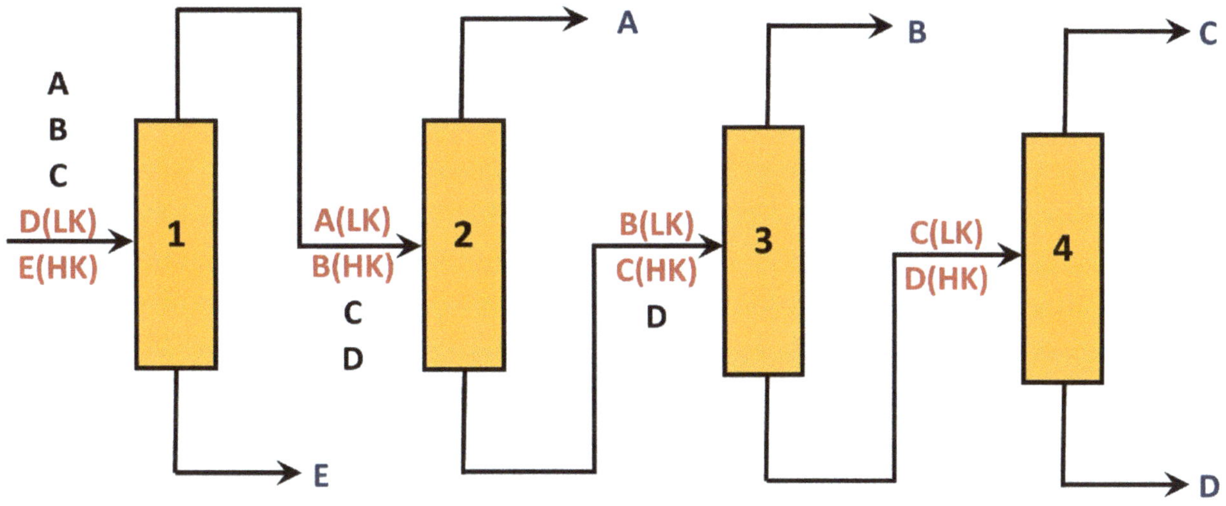
A
B
C
D(LK)
E(HK)
1
E
A(LK)
B(HK)
C
D
2
A
B(LK)
C(HK)
D
3
B
C(LK)
D(HK)
4
C
D

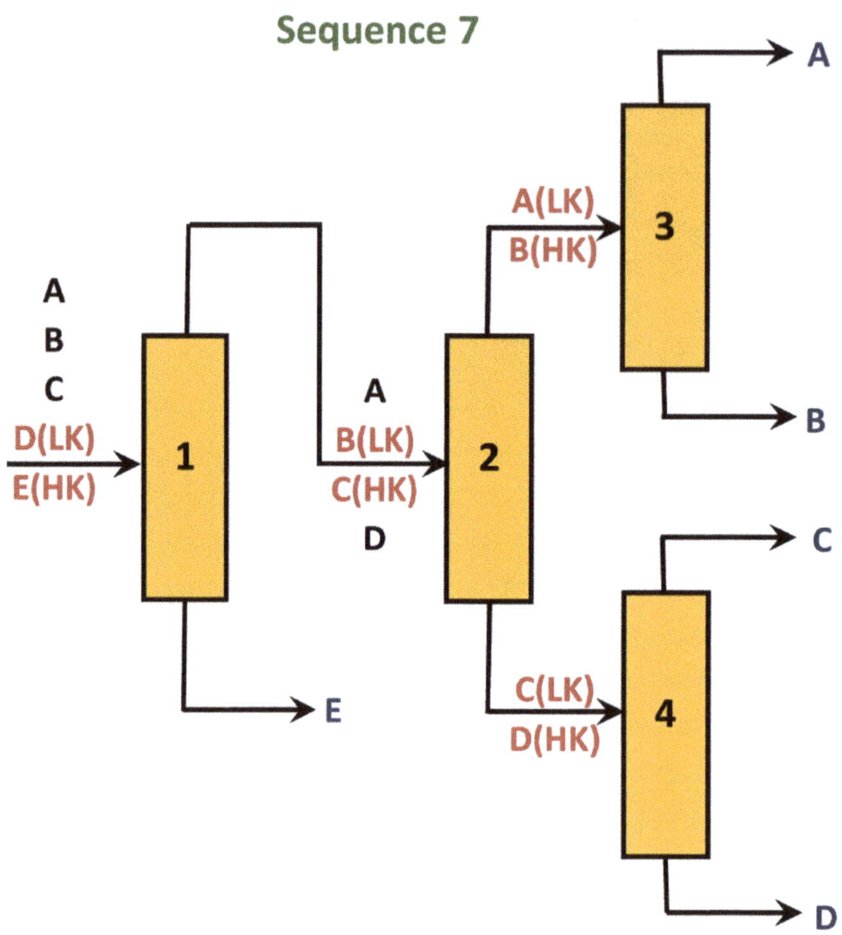
Sequence 7
A
B
C
D(LK)
E(HK)
1
E
A
B(LK)
C(HK)
D
2
A(LK)
B(HK)
3
A
B
C(LK)
D(HK)
4
C
D

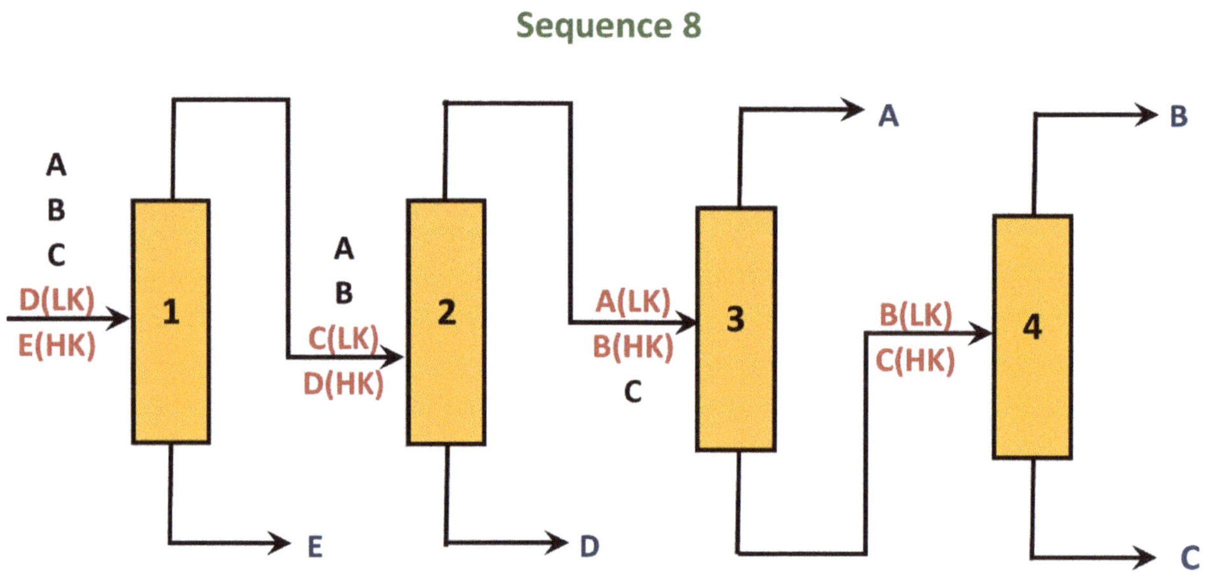
Sequence 8
A
B
C
D(LK)
E(HK)
1
E
A
B
C(LK)
D(HK)
2
D
A(LK)
B(HK)
C
3
A
B(LK)
C(HK)
4
B
C

Sequence 9

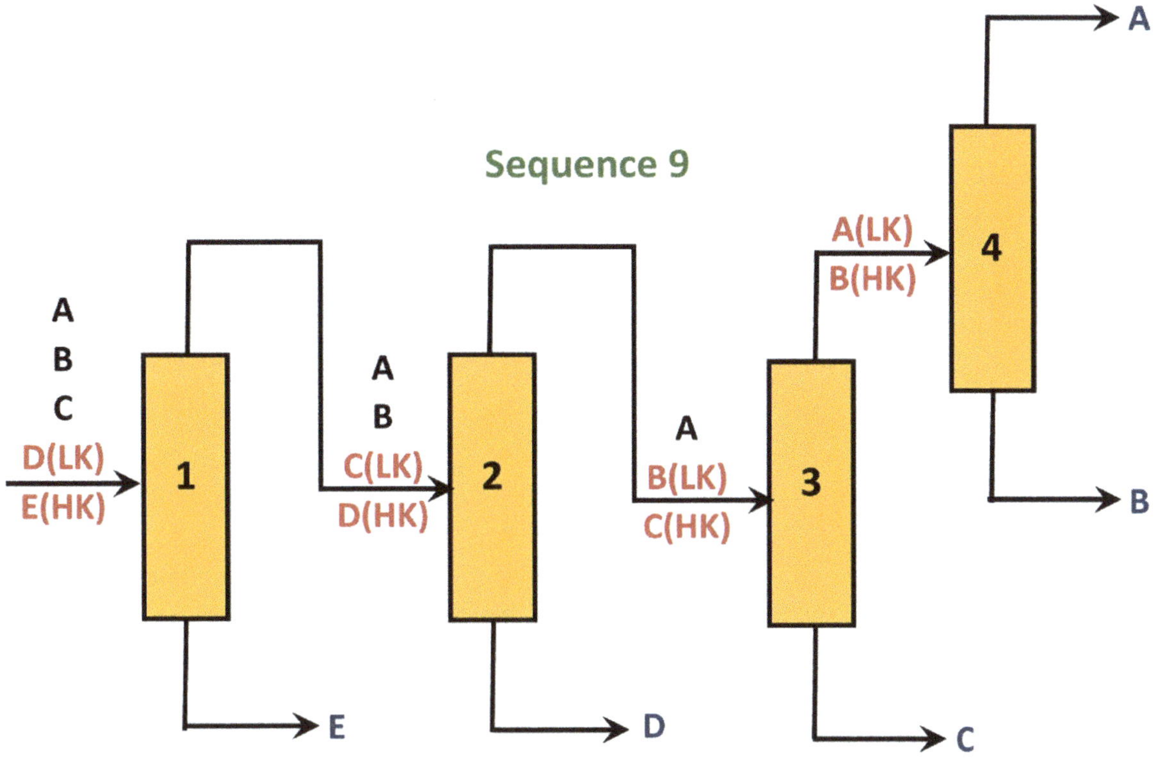

Sequence 10

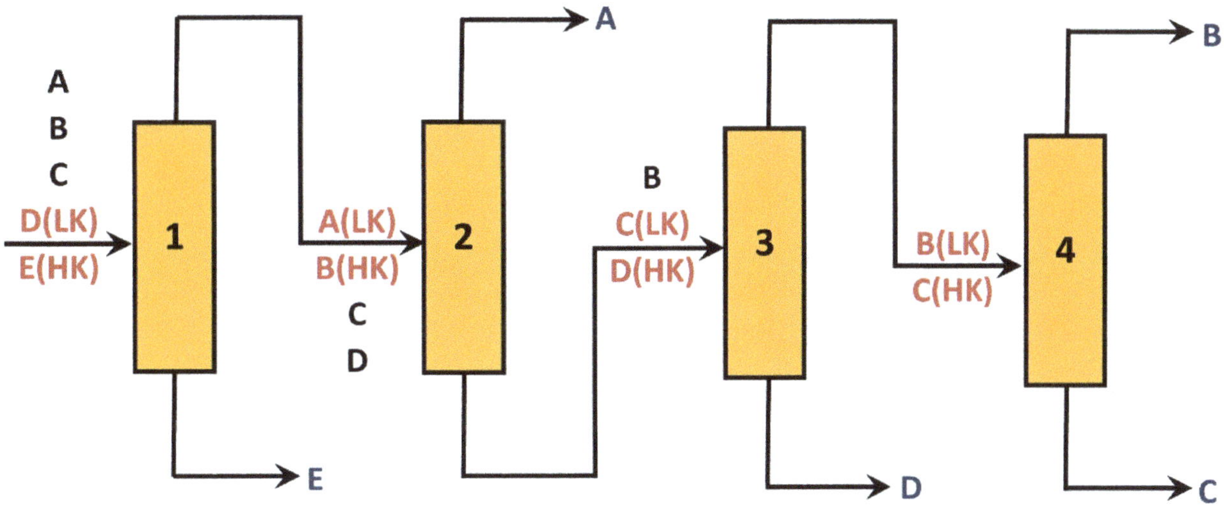

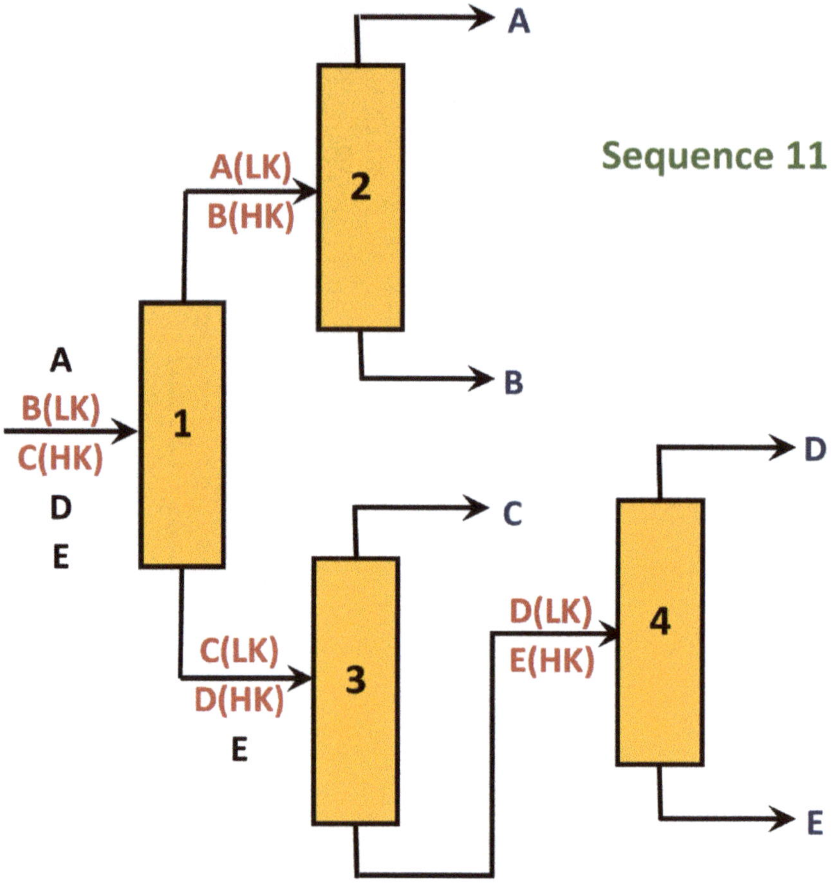
A
Sequence 11
A(LK)
B(HK)
2
B
A
B(LK)
C(HK)
D
E
1
D
C
D(LK)
E(HK)
4
C(LK)
D(HK)
E
3
E

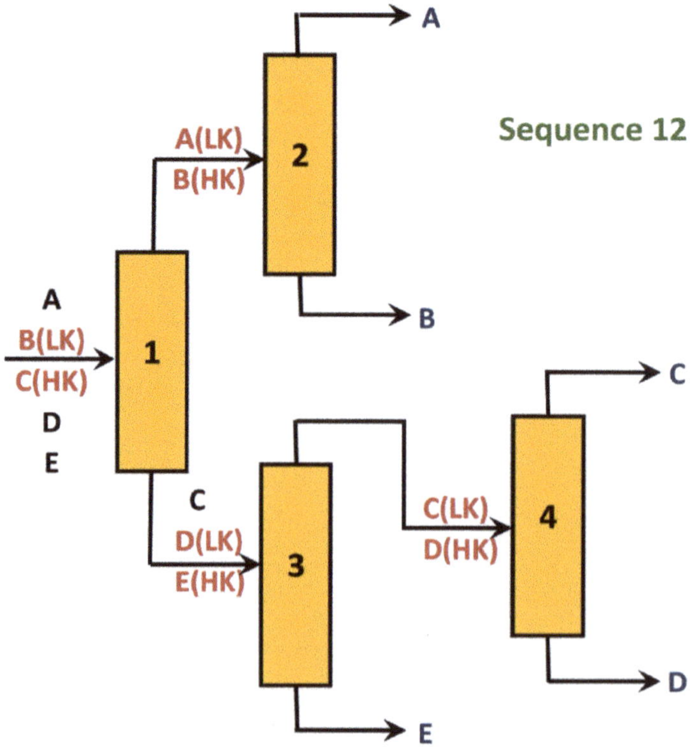
A
Sequence 12
A(LK)
B(HK)
2
B
A
B(LK)
C(HK)
D
E
1
C
C
D(LK)
E(HK)
3
C(LK)
D(HK)
4
D
E

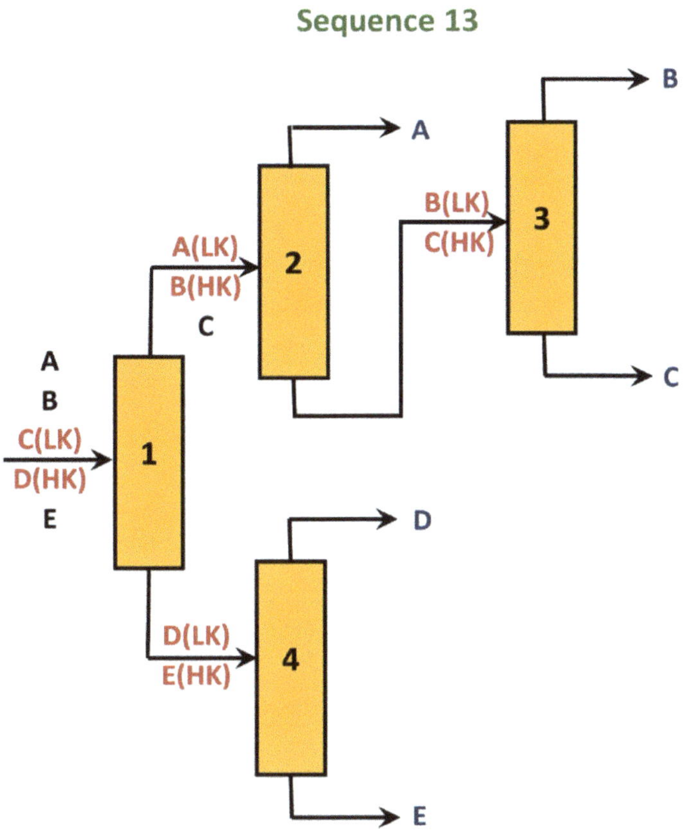
Sequence 13
A
B
C(LK)
D(HK)
E
1
A(LK)
B(HK)
C
2
A
B(LK)
C(HK)
3
B
C
D(LK)
E(HK)
4
D
E

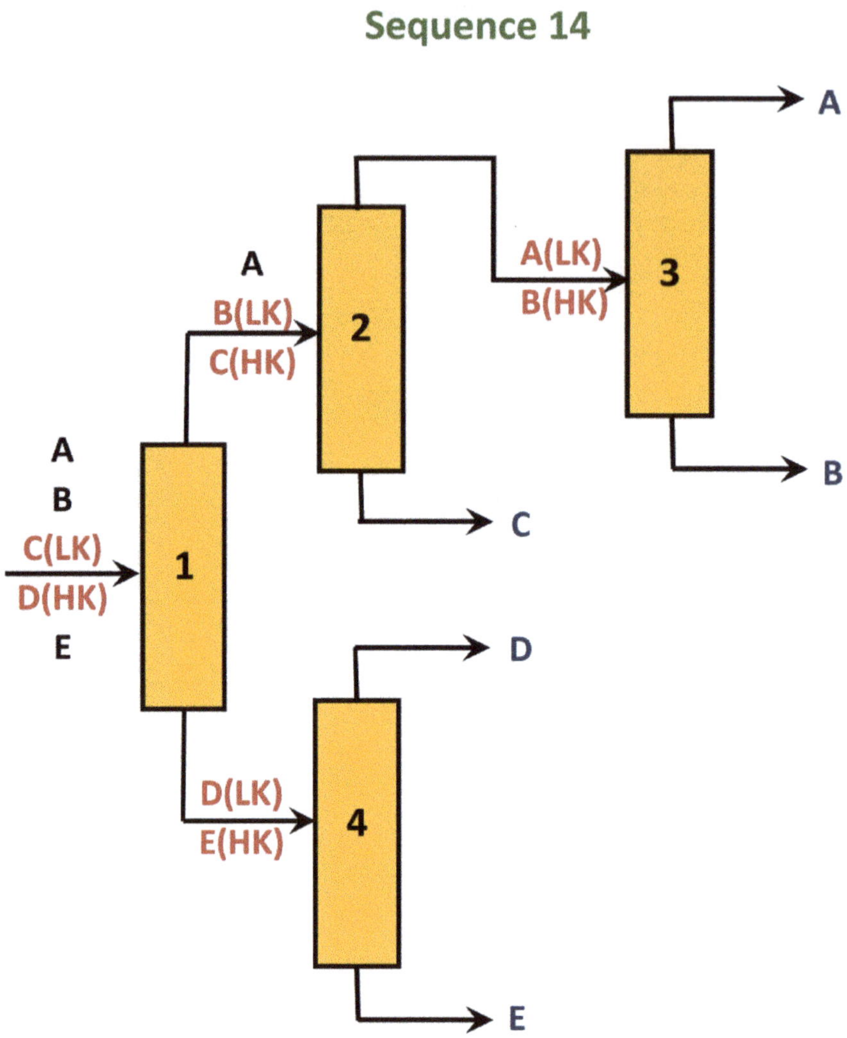
Sequence 14
A
B
C(LK)
D(HK)
E
1
A
B(LK)
C(HK)
2
A(LK)
B(HK)
3
A
B
C
D(LK)
E(HK)
4
D
E

Chapter 14

Liquid-Liquid Extraction

Problem 1

A mixture of two liquids A (solute) and B (diluent) are to be separated using solvent C. The equilibrium data in terms of weight percent for the ternary system A-B-C are given below:

Composition of coexisting liquid phases in equilibrium					
Extract layer			Raffinate layer		
A%	B%	C%	A%	B%	C%
0	0.18	99.82	0	99.89	0.11
5.21	0.32	94.47	5	94.82	0.18
10.79	0.49	88.72	10	89.79	0.21
16.20	0.63	83.17	15	84.78	0.24
22.23	0.79	76.93	20	79.69	0.31
29.01	1.11	69.82	25	74.58	0.42
37.48	1.72	60.80	30	69.42	0.58
43.28	2.33	54.39	35	64.22	0.78
49.44	3.05	47.51	40	58.69	1.36
54.92	4.28	40.80	45	52.76	2.24
59.19	7.24	33.57	50	46.28	3.72
61.79	13.83	24.38	55	38.69	6.31
60.58	25.66	13.76	60.58	25.66	13.76

a) Plot the triangular phase diagram using equilateral-triangle diagram. Show the Plait point.
b) Plot the tie lines on the triangular phase diagram
c) Repeat (a) and (b) using right-angled triangle diagram

Solution

The solutions to parts (a) and (b) are shown in the first figure given below. The solution to part (c) is shown in the second figure given below.

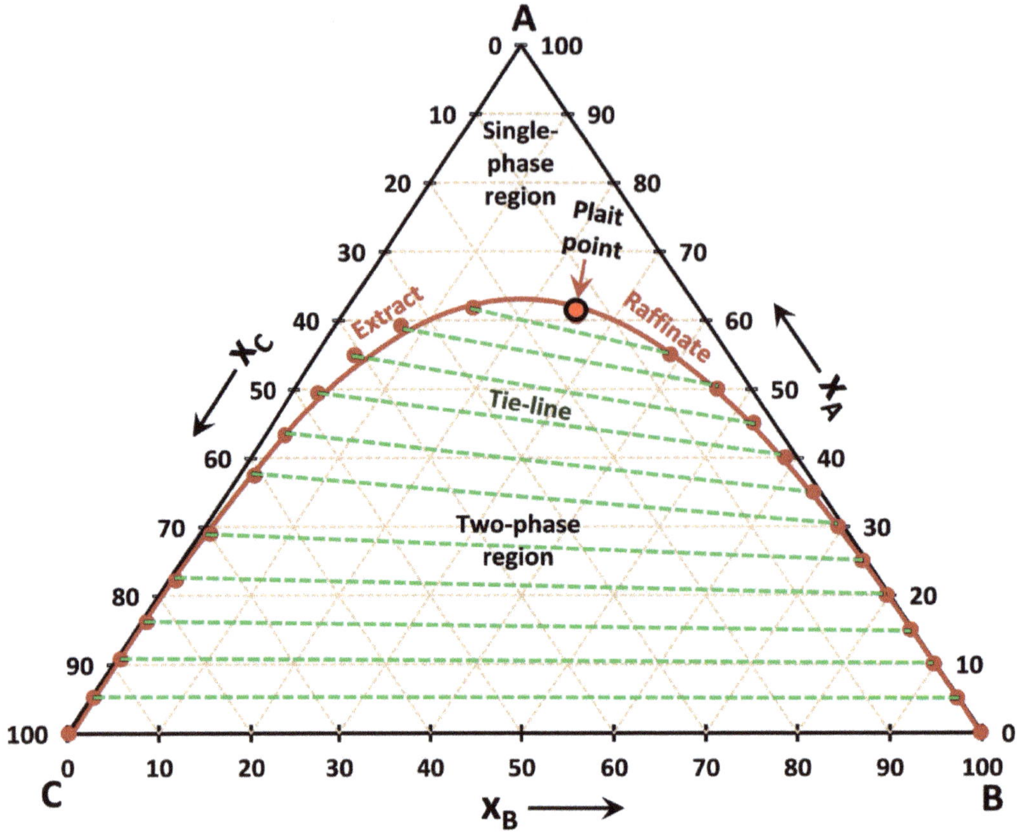
A
B
C
Single-phase region
Plait point
Extract
Raffinate
Tie-line
Two-phase region
x_C
x_A
x_B

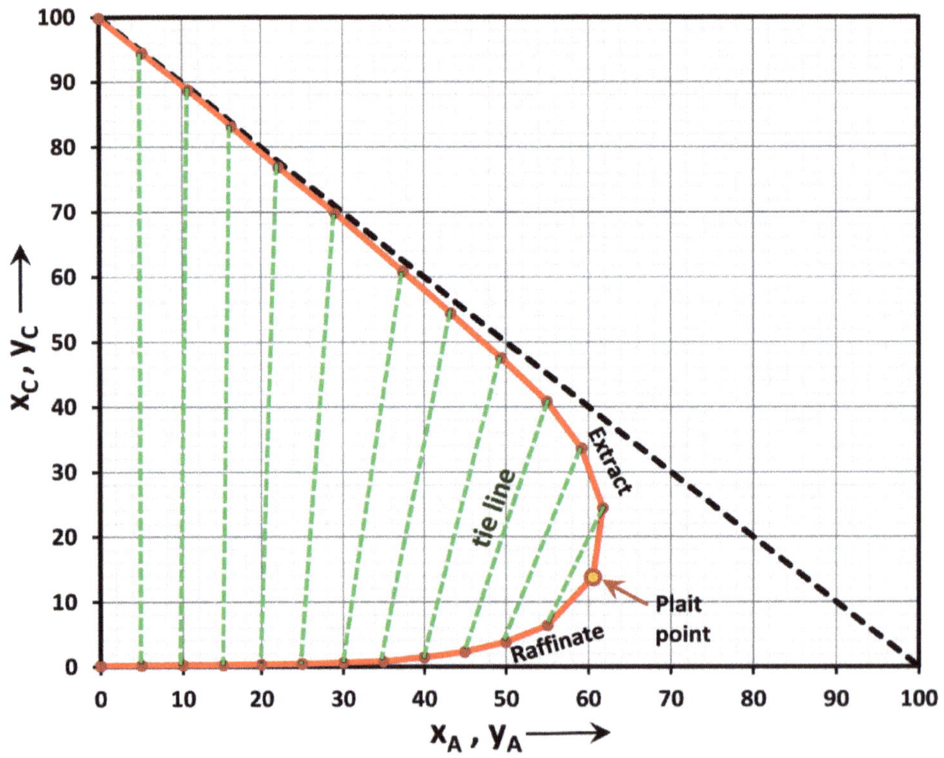
x_C, y_C
x_A, y_A
Extract
tie line
Plait point
Raffinate

Note that the hypotenuses of the right-angled triangle diagram shown above represent the constant diluent (B) concentration lines. The origin of the diagram represents 100% B.

Problem 2

For the ternary system of practice problem 1 given above, a single stage extraction process is used in which 350 kg of a feed solution containing 35 wt% solute A in diluent B is contacted with 350 kg of pure solvent C.
Calculate the amounts and compositions of the extract and raffinate layers. What is the percent recovery of solute A?

Solution

The process is shown schematically below.

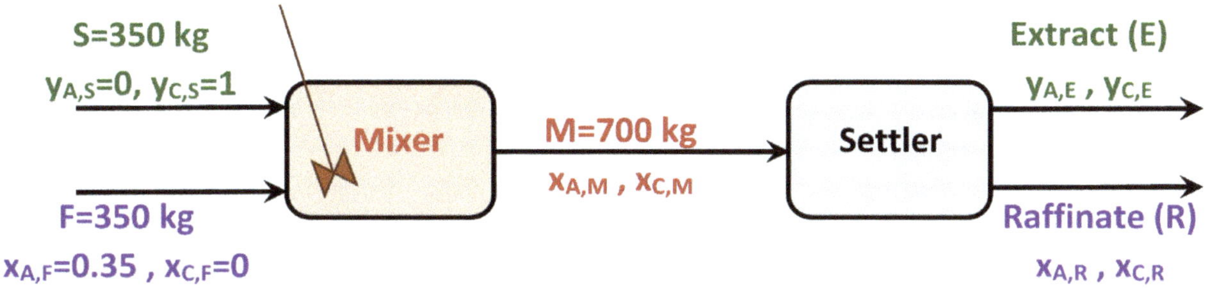

From Equations (14.15) and (14.16), we get the mixture composition:

$$x_{A,M} = \frac{y_{A,S}+x_{A,F}(F/S)}{1+(F/S)} = \frac{0+0.35\times(350/350)}{1+(350/350)} = 0.175$$

$$x_{C,M} = \frac{y_{C,S}+x_{C,F}(F/S)}{1+(F/S)} = \frac{1+0}{1+(350/350)} = 0.50$$

The solution is obtained graphically as shown below. The feed and solvent points are located and connected by a straight line (line SF) on a triangular diagram. The mixture point M falls on the SF line. A tie line passing through the mixture point M is drawn. The compositions of the extract and raffinate are indicated on the triangular diagram by the points E and R, respectively. The compositions are as follows:

Extract composition: $y_{A,E} = 0.18$ and $y_{C,E} = 0.815$

Raffinate composition: $x_{A,R} = 0.165$ and $x_{C,R} = 0.0025$

From Equation (14.19), we get:

$$\frac{E}{M} = \frac{x_{A,M}-x_{A,R}}{y_{A,E}-x_{A,R}} = \frac{0.175-0.165}{0.18-0.165} = 0.667$$

Thus, $E = 0.667 \times 700 = 467$ kg and $R = M - E = 233$ kg.

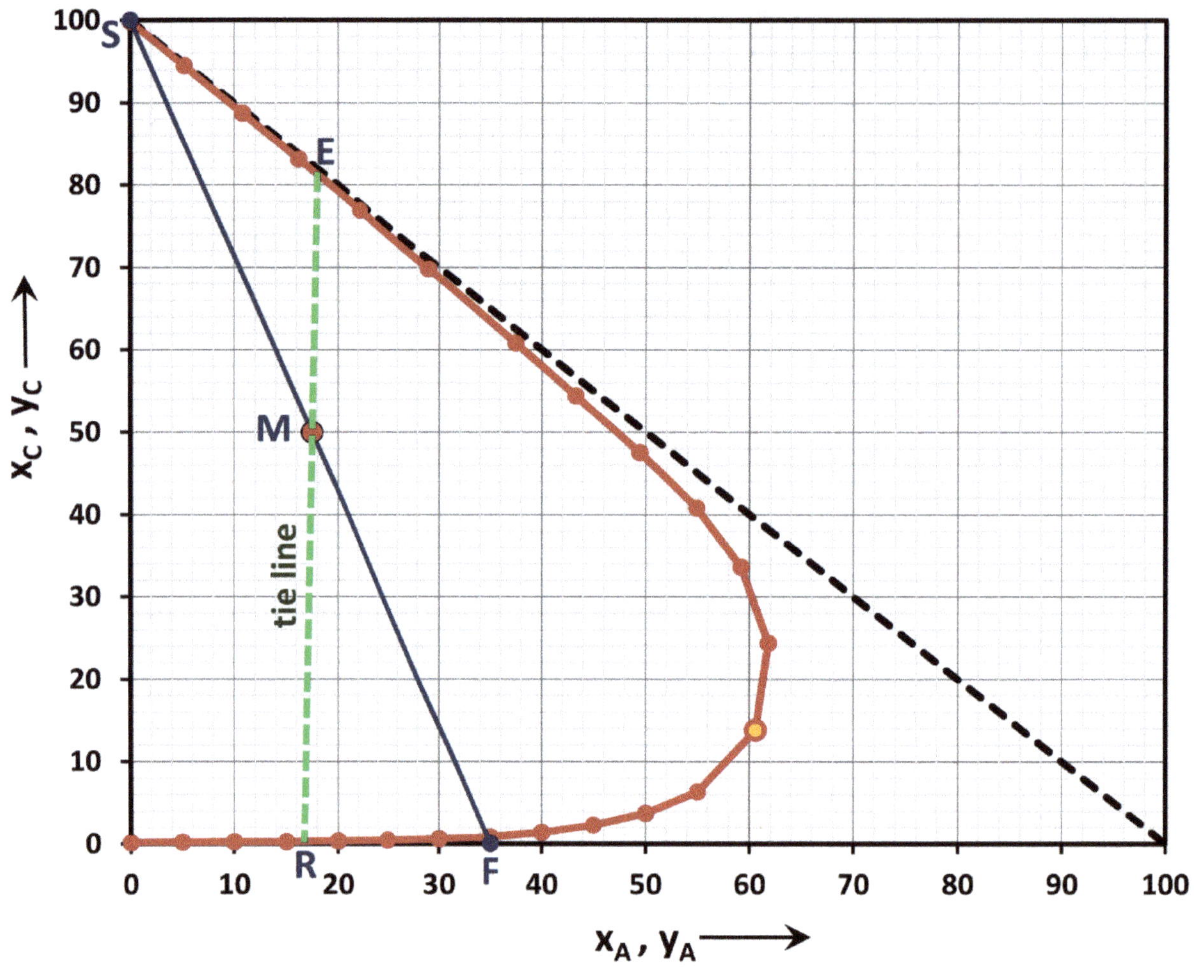

The percent recovery of solute is as follows:

Recovery $= \frac{y_{A,E} \times E}{x_{A,F} \times F} = \frac{0.18 \times 467}{0.35 \times 350}$ =0.686, that is 68.6 %

Problem 3

Pure solvent (species C) is being used to extract a solute (species A) from feed containing 40 wt% A and 60 wt% diluent (species B) using countercurrent multistage extraction. The flow rate of entering solvent is 525 kg/h and the flow rate of feed solution is 185 kg/h. The desired exit solute concentration in the raffinate is 5 wt%.

Estimate the compositions and mass flow rates of extract and raffinate streams leaving the process.

The equilibrium data in terms of weight percent for the ternary system A-B-C are given in practice problem 1 above.

Solution

The procedure to solve this type of problem is discussed in section 14.5.2. A schematic diagram of the process is shown below.

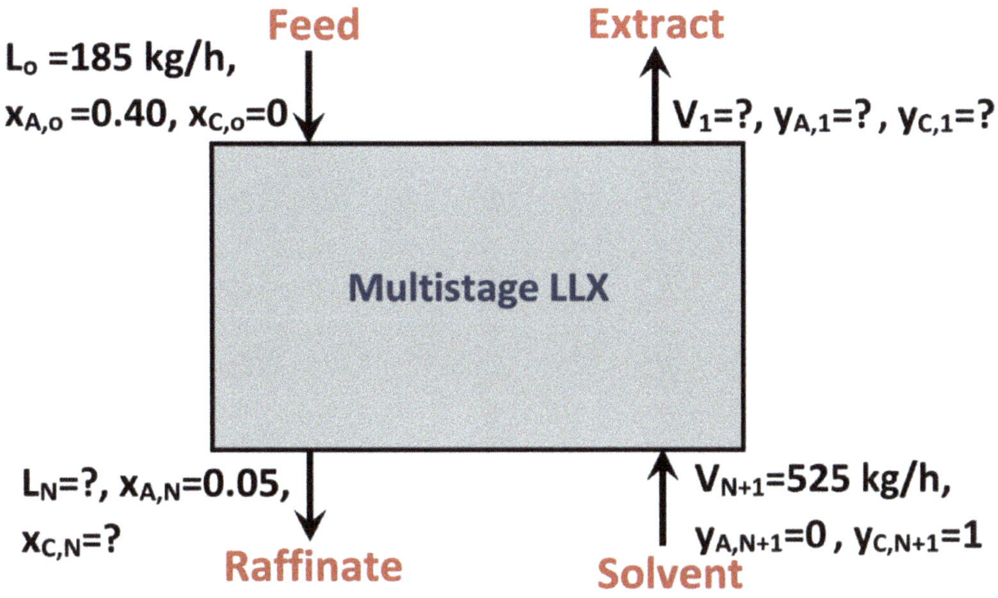

Using Equations (14.32) and (14.33), we obtain the mixture (M) composition as: $x_{A,M} = 0.1042$, $x_{C,M} = 0.7394$. The points V_{N+1}, L_o and M are located on the triangular diagram as shown below. The coordinates of these points are as follows: $V_{N+1}(y_{A,N+1} = 0,\ y_{C,N+1} = 1.0)$; $L_o(x_{A,o} = 0.40,\ x_{C,o} = 0)$; $M(x_{A,M} = 0.1042,\ x_{C,M} = 0.7394)$. Note that the points V_{N+1}, L_o and M are collinear. The point L_N is located next. This point falls on the equilibrium curve (raffinate portion) at $x_{A,N} = 0.05$. The intersection of the line $L_N M$ with the equilibrium curve (extract portion) locates the point V_1. From the triangular diagram, the compositions of extract and raffinate streams are as follows: Extract:$y_{A,1} = 0.116$; $y_{C,1} = 0.88$; Raffinate: $x_{A,N} = 0.05$; $x_{C,N} = 0.0018$. Note that the triangular diagram shows percent composition (not mass fractions).

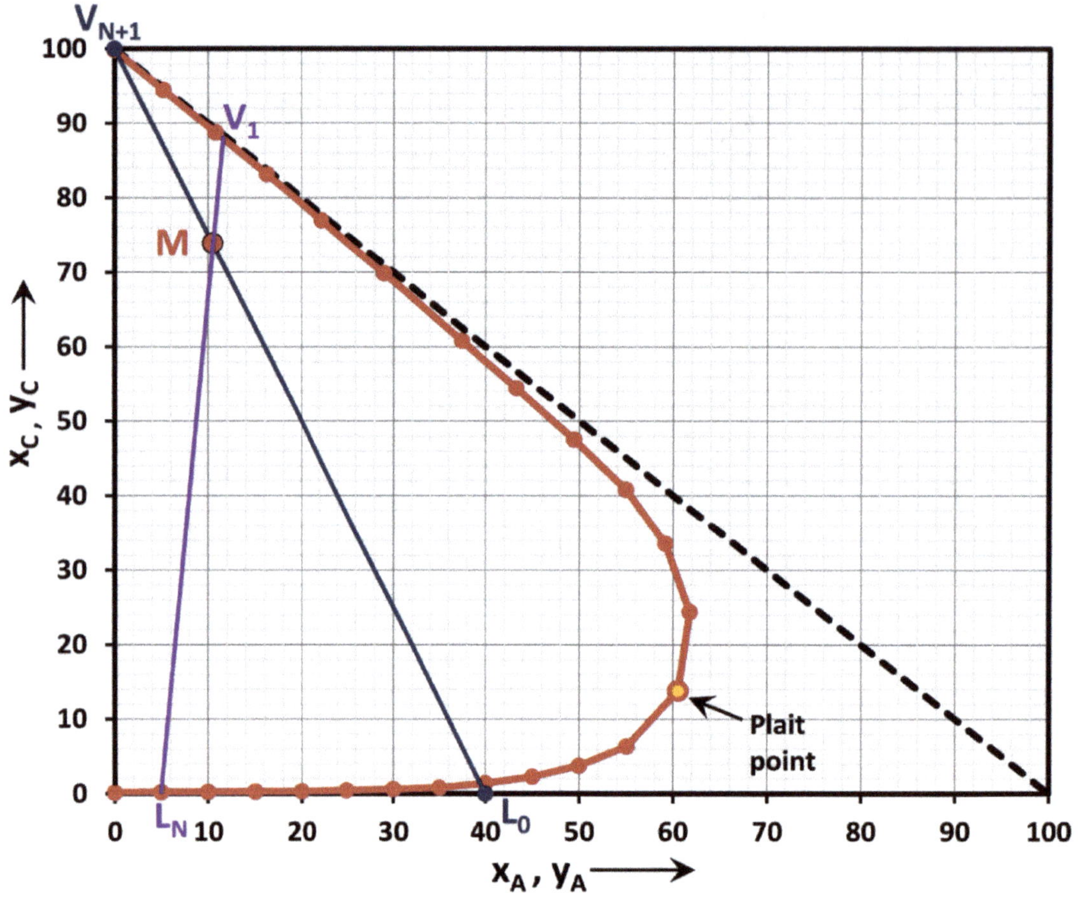

From Equation (14.36), we get:

$$\frac{V_1}{M} = \frac{x_{A,M} - x_{A,N}}{y_{A,1} - x_{A,N}} = \frac{0.1042 - 0.05}{0.116 - 0.05} = 0.8212$$

As $M = L_o + V_{N+1} = 185 + 525 = 710$ kg/h, it follows that: $V_1 = 0.8212 \times 710 = 583.05$ kg/h and $L_N = M - V_1 = 710 - 583.05 = 126.95$ kg/h.

Problem 4

A single stage liquid-liquid extraction process is being used to recover an antibiotic from dilute aqueous fermentation broth containing 0.75 wt% antibiotic using pure solvent. The ratio of solvent mass flow rate to feed mass flow rate is 0.25 ($V/L = 0.25$). The equilibrium relation is $y = K_d x$ where K_d=85 (y and x are mass fractions of antibiotic in extract and reffinate phases, respectively at equilibrium).
What are the compositions of the exit extract and raffinate streams?
What is the percent recovery of antibiotic?
You can assume that solvent is insoluble in water.

Solution

The process is shown schematically below.

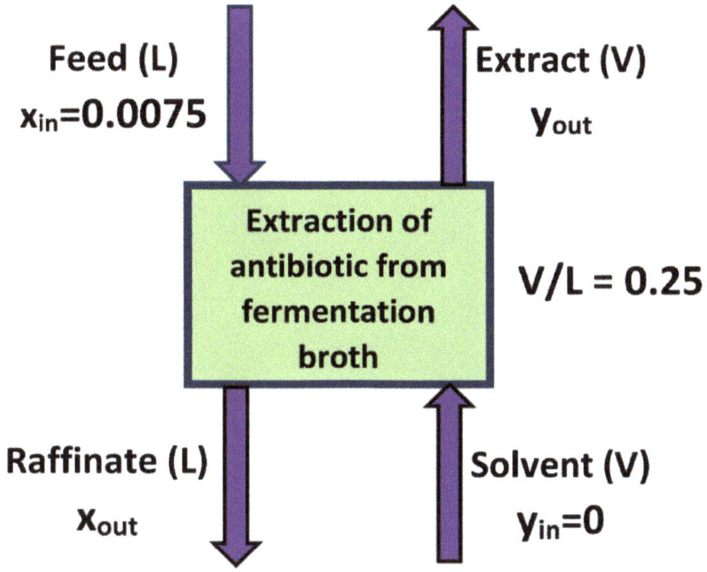

The extraction factor, defined in Equation (14.10), for the process is:

$$E_o = \frac{K_d}{L/V} = 85 \times 0.25 = 21.25$$

Equation (14.9) gives

$$\frac{x_{out}}{x_{in}} = \left(\frac{1}{1+E_o}\right)\left[1 + \frac{V}{L}\left(\frac{y_{in}}{x_{in}}\right)\right] = \frac{1}{1+21.25} = 0.04494$$

$$x_{out} = 0.04494 \times 0.0075 = 3.3705 \times 10^{-4}$$

The extract composition is

$$y_{out} = 85 \times 3.3705 \times 10^{-4} = 0.02865$$

The percent recovery of antibiotic is:

$$\%\,Recovery\ of\ antibiotic = 100\left(1 - \frac{x_{out}}{x_{in}}\right) = 100(1 - 0.04494) = 95.506$$

Excellent recovery in a single stage as the extraction factor is high.

Problem 5

A single stage liquid-liquid extraction process is being used to recover solute A from dilute aqueous feed stream containing 1 wt% solute. The feed mass flow rate is 1000 kg/h. Pure solvent C is used to extract the solute.

If the concentration of solute in the raffinate is desired to be 0.1 wt%, estimate the mass flow rate of solvent.

The equilibrium relation is $y = K_d x$ where K_d=0.75 (y and x are mass fractions of solute in extract and reffinate phases, respectively at equilibrium). You can assume that solvent is insoluble in water.

Solution

The process is shown schematically below.

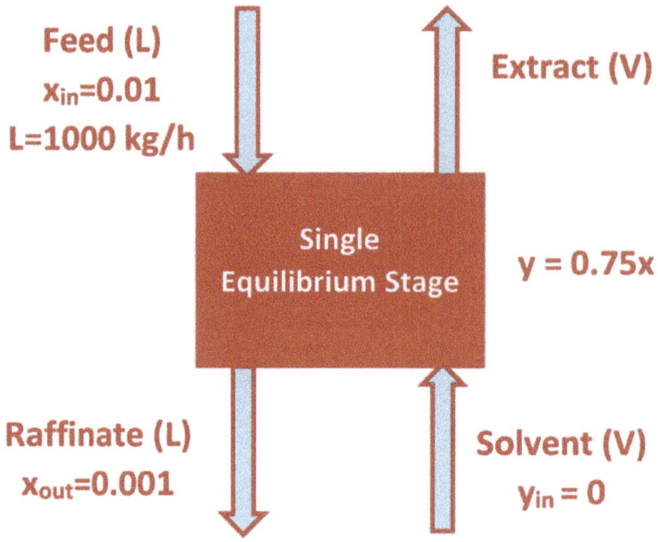

Equation (14.9) gives

$$\frac{x_{out}}{x_{in}} = \left(\frac{1}{1+E_o}\right)\left[1 + \frac{V}{L}\left(\frac{y_{in}}{x_{in}}\right)\right] = \left(\frac{1}{1+E_o}\right)[1+0] = \left(\frac{1}{1+E_o}\right)$$

$$E_o = \frac{x_{in}}{x_{out}} - 1 = \frac{0.01}{0.001} - 1 = 9$$

$$E_o = \frac{K_d}{L/V} = 9$$

$$\frac{L}{V} = \frac{K_d}{E_o} = \frac{0.75}{9} = 0.0833$$

$$V = \frac{L}{0.0833} = \frac{1000}{0.0833} = 12{,}004.8 \text{ kg/h}$$

The mass flow rate of solvent required is approximately 12 times that of the feed flow rate. One reason for high solvent flow rate requirement is a small value of the distribution coefficient K_d . A small value of K_d is indicative of low solubility of solute in the solvent phase.

Problem 6

A multistage liquid-liquid extraction process is being used to remove solute from feed stream using pure solvent C. The feed mass flow rate is 50 kg/h and the solvent mass flow rate is 100 kg/h. The mass fraction of solute in the feed stream is 0.01.
Estimate the number of equilibrium stages required to reduce the concentration of solute in the reffinate stream leaving the process to 0.001.
The equilibrium relation is $y = K_d x$ where K_d=0.85 (y and x are mass fractions of solute in extract and reffinate phases, respectively at equilibrium).
You can use Kremser equation to estimate the number of stages.

Solution

The process is shown schematically below.

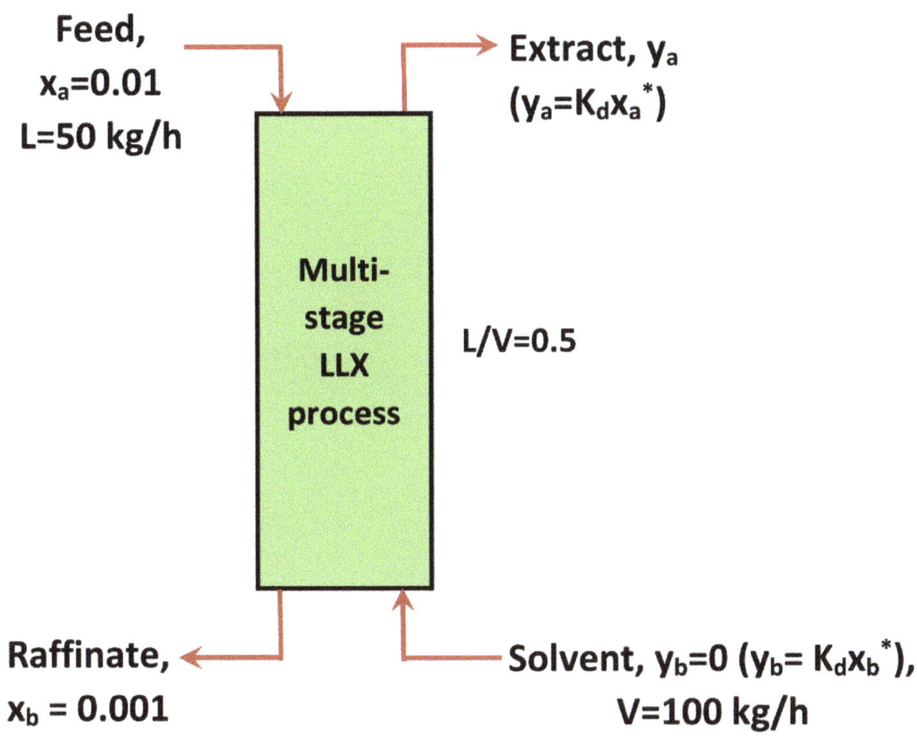

The solute balance gives:

$$y_a = y_b - \left(\frac{L}{V}\right)(x_b - x_a) = 0 - 0.50 \times (0.001 - 0.01) = 0.0045$$

The extraction factor is:

$$E_o = \frac{K_d}{L/V} = \frac{0.85}{0.5} = 1.70$$

The following Kremser equation, Equation (14.27), can now be used to calculate the number of stages required:

$$N = \frac{\ln\left(\frac{x_a - x_a^*}{x_b - x_b^*}\right)}{\ln(E_o)} \qquad when\ E_o \neq 1$$

where $x_a^* = y_a/K_d = \frac{0.0045}{0.85} = 5.294 \times 10^{-3}$ and $x_b^* = y_b/K_d = 0$. The number of stages turn out to be:

$$N = \frac{\ln\left(\frac{0.01 - 5.294 \times 10^{-3}}{0.001 - 0}\right)}{\ln(1.70)} = 2.92$$

Thus approximately 3 stages are needed to reduce the concentration (mass fraction) of solute in the raffinate stream to 0.001.

Chapter 15

Solid-Liquid Leaching

Problem 1

A single stage solid-liquid leaching process is being used to extract oil from flaked oil beans with a suitable solvent. 750 kg of oil beans is leached with 650 kg of pure solvent. The oil content of the oil beans is 21 wt%. At equilibrium, N is approximately 2.1 kg inert solid/kg solution retained for the underflow slurry and $N = 0$ for the overflow stream. The tie lines are vertical lines expressed as $y = x$.
Calculate the amounts and compositions of the exit leachate and leached solids slurry.

Solution

Refer to the schematic diagram shown below.

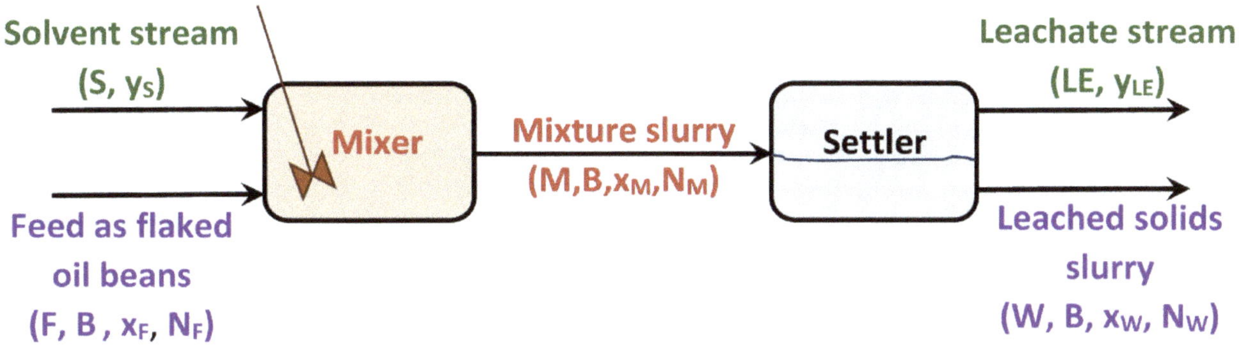

$$B = 0.79 \times 750 = 592.5\ kg \text{ (inert solids)}$$

$$F = 0.21 \times 750 = 157.5\ kg \text{ (Feed solution, 100\% solute as } x_F = 1)$$

$$S = 650\ kg \text{ (pure solvent, } y_S = 0)$$

$$B = N_F F; \quad N_F = B/F = 592.5/157.5 = 3.762$$

$$F + S = M = 650 + 157.5 = 807.5\ kg$$

$$x_M = \frac{y_S + x_F(F/S)}{1 + (F/S)} = \frac{0 + 1(157.5/650)}{1 + (157.5/650)} = 0.195$$

$$B = N_M M; \; N_M = B/M = 592.5/807.5 = 0.734$$

We can now obtain the solution graphically as shown in the figure below.

- Locate the feed point $\boldsymbol{F}(x_F = 1, N_F = 3.762)$ on the phase diagram
- Locate the solvent point $\boldsymbol{S}(y_S = 0, N_S = 0)$ on the phase diagram
- Locate the mixture point $\boldsymbol{M}(x_M = 0.195, N_M = 0.734)$. $\boldsymbol{M}$ lies on the FS line

The vertical tie line drawn through the mixture point ***M*** gives the following results:

$$x_W = 0.195; \quad N_W = 2.1; \quad y_{LE} = 0.195; \quad N_{LE} = 0$$

Thus,

$$W = B/N_W = 592.5/2.1 = 282.143\ kg$$

$$LE = M - W = 807.5 - 282.143 = 525.357\ kg$$

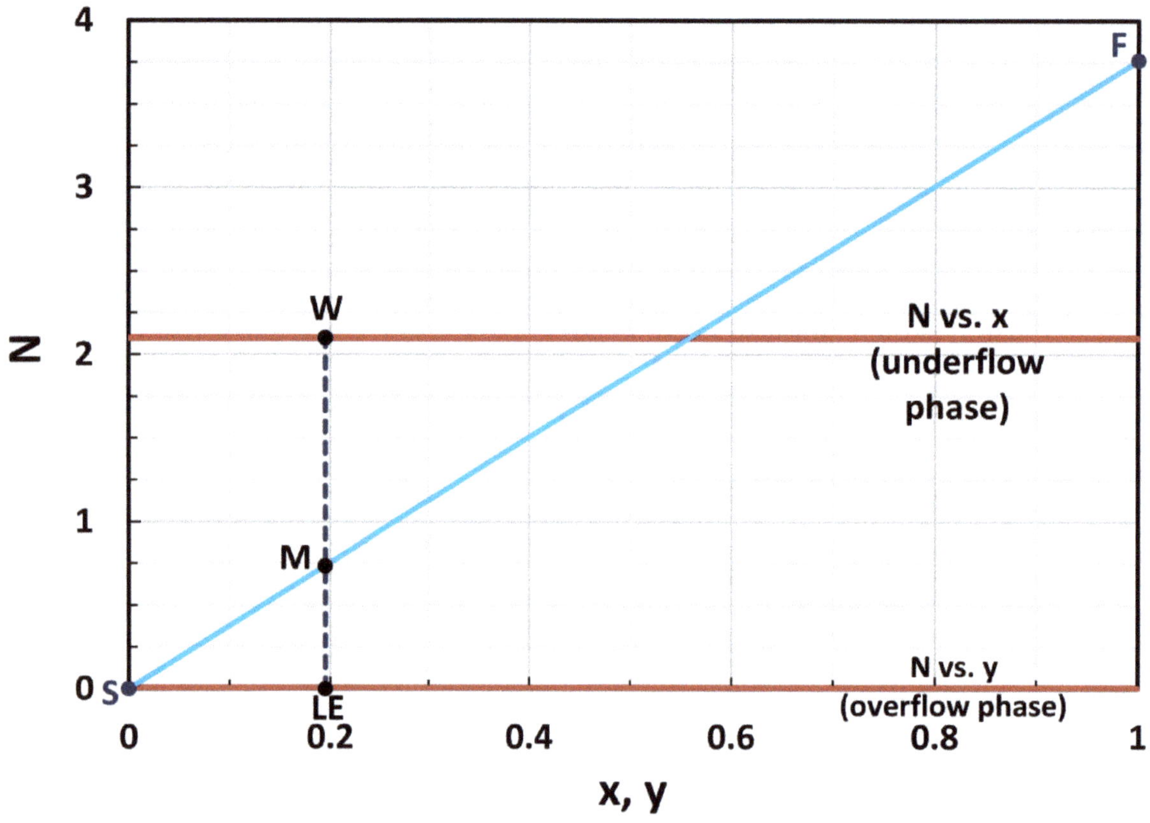

Problem 2

We wish to leach oil from meal by a countercurrent multistage leaching process using a pure solvent. The feed flow rate is 1000 kg/h of meal (inert solids +oil). The oil content of the meal is 19 wt%. The pure solvent flow rate entering the process is 700 kg/h. The desired mass fraction of oil in the underflow product solution is 0.075. Estimate the number of stages required. What are the flow rates and compositions of the streams leaving the process?
The equilibrium data are as follows:

y	0	0.1	0.2	0.3	0.4	0.5	0.6	0.7
N	2.10	2.08	2.04	1.99	1.92	1.85	1.78	1.71

Note that $N = 0$ for the overflow stream at equilibrium and the tie lines are vertical lines expressed as $y = x$.

Solution

The procedure to solve multistage leaching problem is discussed in section 15.5.1. We will first calculate the leached slurry (underflow) and exit leachate (overflow) mass flow rates (L_N, V_1) and the unknown (N_N, y_1), given the feed and solvent flow rates and compositions: $L_o, V_{N+1}, x_o, y_{N+1}$. The quantities specified are shown in the figure below.

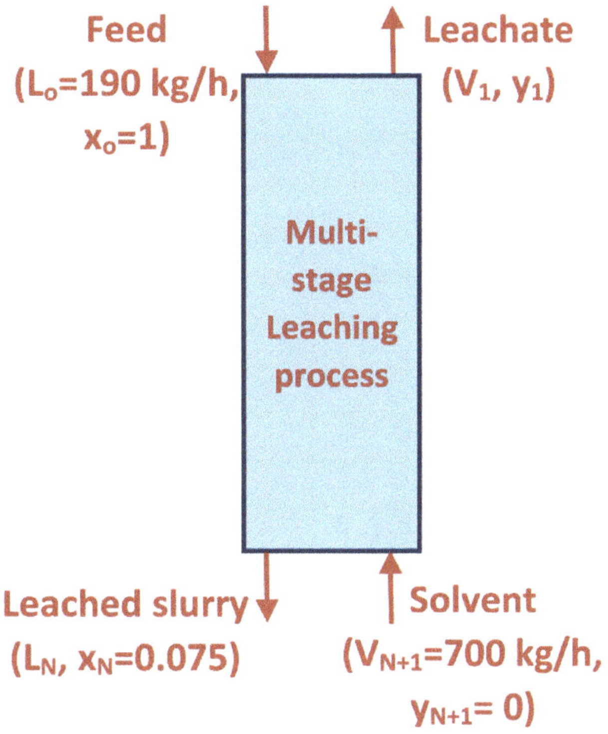

$$L_o = 0.19 \times 1000 = 190\ kg/h$$

$$B = 0.81 \times 1000 = 810\ kg/h$$

$$B = N_o L_o; \quad N_o = B/L_o = 810/190 = 4.263$$

$$M = V_{N+1} + L_o = 700 + 190 = 890\ kg/h$$

$$x_M = \frac{y_{N+1} + x_o(L_o/V_{N+1})}{1 + (L_o/V_{N+1})} = \frac{0 + 1(190/700)}{1 + (190/700)} = 0.2135$$

The unknowns (N_N, y_1), obtained graphically as shown in the figure below, are:

$$N_N = 2.085; \ y_1 = 0.32$$

Thus,

$$L_N = B/N_N = 810/2.085 = 388.49\ \ kg/h$$

$$V_1 = M - L_N = 890 - 388.49 = 501.51\ kg/h$$

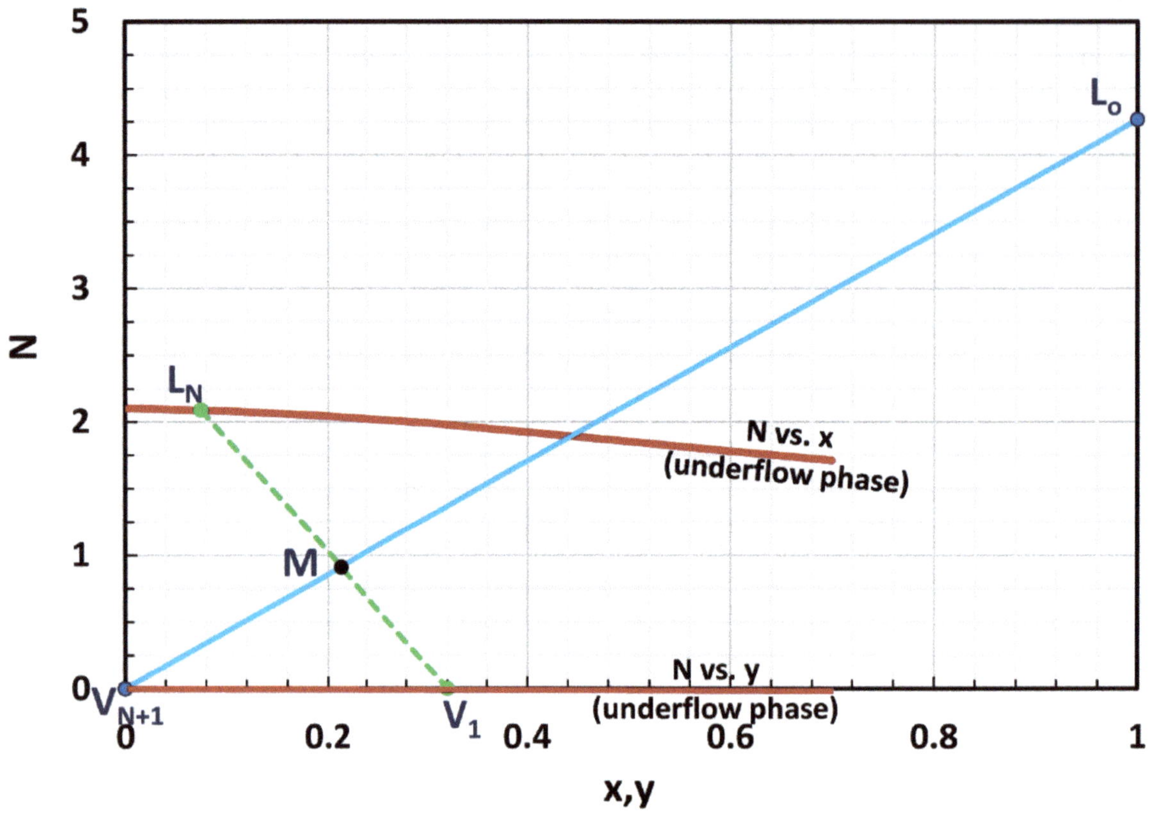

Let us now estimate the number of stages required. The difference point coordinates are:

$$x_{diff} = \frac{x_o L_o - y_1 V_1}{L_o - V_1} = \frac{(1 \times 190) - (0.32 \times 501.51)}{190 - 501.51} = -0.095$$

$$N_{diff} = \frac{B}{L_o - V_1} = \frac{810}{190 - 501.51} = -2.6$$

Given the difference point coordinates ($x_{diff} = -0.095, N_{diff} = -2.6$) and feed point coordinates $L_o(x_o = 1, N_o = 4.263)$, we can estimate the number of stages graphically as shown in the figure below. Note that the point L_o is not shown as it falls outside the figure. Also the coordinates of the point L_N are ($x_N = 0.075, N_N = 2.085$).

The number of stages required is about 2.5.

The flow rates and compositions of the streams leaving the process are:

$$V_1 = 501.51\ kg/h, \quad y_1 = 0.32$$

$$L_N = 388.49\ \ kg/h\,, \quad x_N = 0.075$$

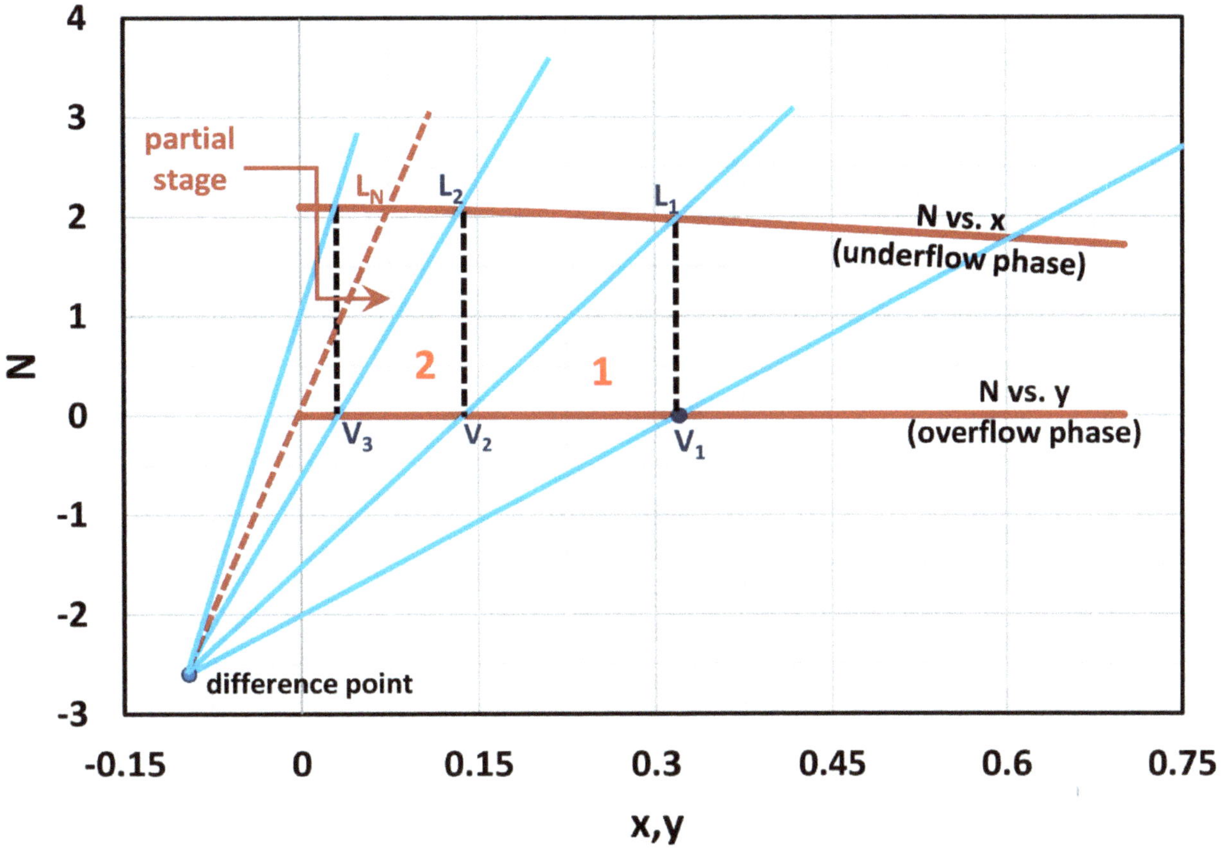

Problem 3

A multi-stage countercurrent leaching process is being used to extract solute from the solid phase with the help of a suitable solvent. The process requires 7 equilibrium stages. The solute concentration of solids entering as feed is 3 wt%. The ratio of liquid mass flow rate to solids mass flow rate is 1.35. The equilibrium relation is $y = 1.25x$. Determine the concentrations of solute in the discharge streams assuming that the solvent entering the process is pure. What is the percent recovery of solute in the solvent?

Solution

Refer to the schematic diagram shown below. From the given information, the leaching factor E_L can be calculated as:

$$E_L = \frac{K}{L/V} = 1.25 \times 1.35 = 1.6875$$

The leaching factor can also be expressed as (see Equation (15.37)):

$$E_L = \frac{x_a - x_b}{x_a^* - x_b^*} = \frac{0.03 - x_b}{x_a^* - 0} = 1.6875$$

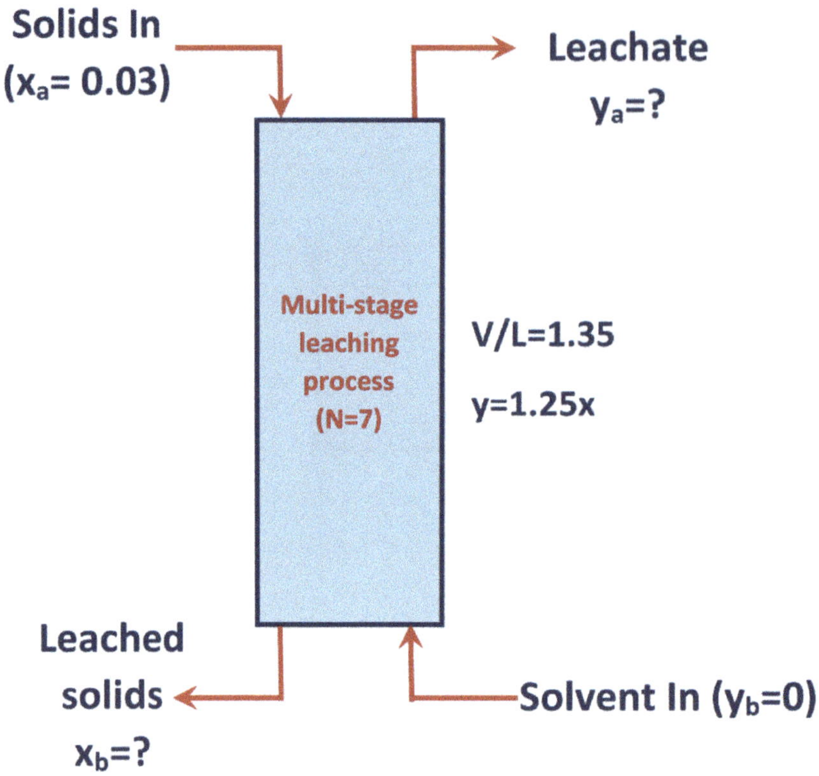

Note that $x_a = 0.03$, $y_b = 0$ and $x_b^* = y_b/K = 0$. Upon rearrangement the above equation becomes:

$$0.03 - x_b = 1.6875 x_a^*$$

The number of stages is $N = 7$. We can now use the Kremser equation (Equation (15.35)) as:

$$N = \frac{\ln\left(\frac{x_a - x_a^*}{x_b - x_b^*}\right)}{\ln(E_L)} = \frac{\ln\left(\frac{0.03 - x_a^*}{x_b - 0}\right)}{\ln(1.6875)} = 7$$

Upon rearrangement, the Kremser equation gives:

$$0.03 - x_a^* = 38.968 x_b$$

Thus we have two equations and two unknowns (x_b *and* x_a^*). We can solve for the unknowns and get the following values:

$$x_b = 3.169 \times 10^{-4} \quad \text{and} \quad x_a^* = 0.01759$$

Now we can calculate the concentration of solute in the exit leachate stream as:

$$y_a = 1.25 x_a^* = 1.25 \times 0.01759 = 0.022$$

The percent recovery of solute in the solvent stream is:

$$\%\, Recovery = \frac{y_a V}{x_a L} \times 100 = \frac{0.022 \times 1.35}{0.03} \times 100 = 99$$

Almost full recovery of solute is achieved using 7 stages.

Chapter 16

Analysis and Design of Continuous Contacting Packed Columns

Problem 1

A packed tower is to be used to absorb SO_2 from air stream using pure water as a solvent. The gas entering the tower contains 2.0 mol% A and 85 percent of it is to be absorbed by water. The molar flow rates of gas and liquid streams are 0.05 kmol/s and 1.98 kmol/s, respectively. Under the tower conditions, the equilibrium relationship is as follows: $y = 35x$. The column height is 4 m and its diameter is 1.25 m. Estimate the value of the mass transfer capacity coefficient $K_y a$. You can assume that the gas and liquid molar flow rates are constant.

Solution

Refer to the schematic diagram of the packed tower shown below.

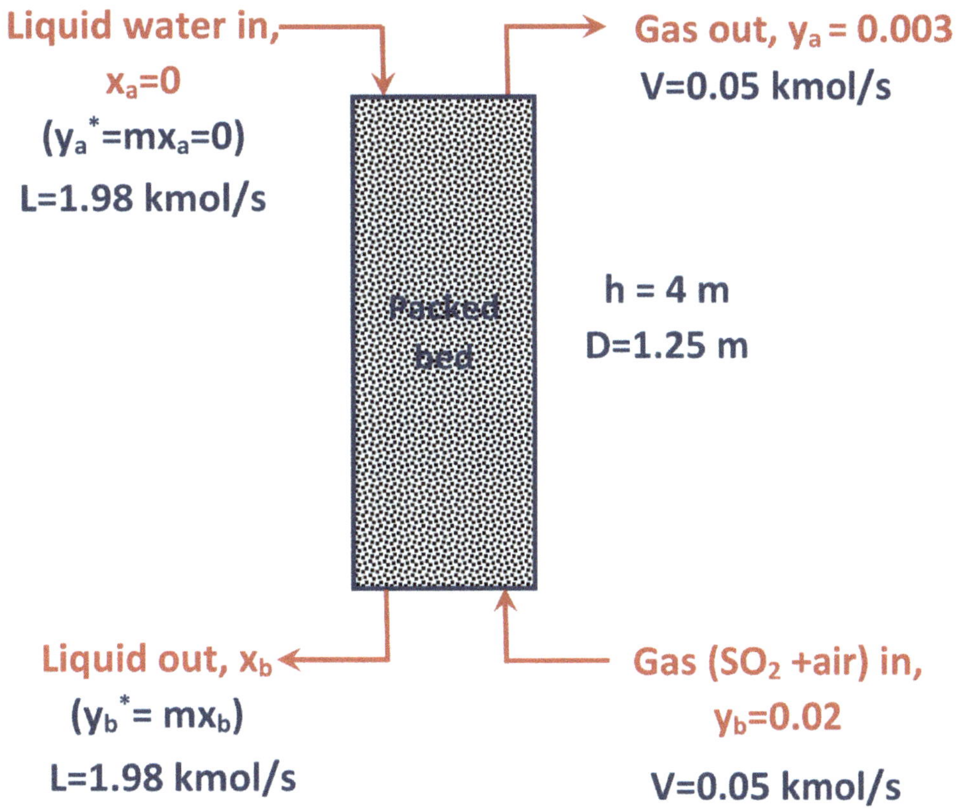

We first need to calculate x_b. From Equation (16.27):

$$y_a = y_b + (L/V)(x_a - x_b)$$

Upon rearrangement,

$$x_b = x_a + (y_b - y_a)/(L/V) = 0 + (0.02 - 0.003)/(1.98/0.05) = 4.293 \times 10^{-4}$$

Note that y_a is 15% of y_b , that is, $y_a = 0.15 \times 0.02 = 0.003$

Thus all the mole fractions at inlet and outlet of the tower are now known.

For linear operating and equilibrium lines with different slopes, we can use Equation (16.28) to determine the NTUs:

$$NTU = \frac{y_b - y_a}{(y - y^*)_{log-mean}} = \frac{0.02 - 0.003}{(y - y^*)_{log-mean}}$$

$$(y - y^*)_{log-mean} = \frac{(y_b - y_b^*) - (y_a - y_a^*)}{ln\left(\frac{y_b - y_b^*}{y_a - y_a^*}\right)} = \frac{(0.02 - 0.015025) - (0.003 - 0)}{ln\left(\frac{0.02 - 0.015025}{0.003 - 0}\right)} = 3.905 \times 10^{-3}$$

$$NTU = \frac{0.02 - 0.003}{3.905 \times 10^{-3}} = 4.3534$$

$$h = (HTU) \times (NTU)$$

$$h = \left(\frac{V/S}{K_y a}\right) \times (NTU)$$

$$K_y a = \left(\frac{V/S}{h}\right) \times (NTU) = \left(\frac{4V}{\pi D^2 h}\right) \times (NTU)$$

$$K_y a = \left(\frac{V/S}{h}\right) \times (NTU) = \left(\frac{4 \times 0.05}{\pi \times (1.25)^2 \times 4}\right) \times (4.3534) = 0.04434\ \frac{kmol}{s.m^3}$$

Problem 2

A packed tower is to be used to absorb cyclohexane from airstream. A non-volatile oil is used to absorb cyclohexane. The air entering the tower contains 1.5 mol% cyclohexane and 90 percent of it is absorbed by oil. The oil entering the column is cyclohexane-free. The tower is operated at 30 °C and 1 atm. The gas and liquid molar flow rates on a unit area basis are 0.5 lb-mol/min.ft^2 and 0.17 lb-mol/min.ft^2, respectively. The equilibrium and mass transfer coefficient data are as follows:

$$y^* = 0.15x\ ;\ k_x a = 0.30\ lb-mol/h.ft^3\ ;\ k_y a = 14.0\ lb-mol/h.ft^3$$

Determine the following:

a) the number of transfer units, NTUs
b) the height of the transfer unit, HTU
c) the height of the packed tower

You can assume that the gas and liquid molar flow rates are constant.

Solution

Refer to the schematic diagram of the packed tower shown below.

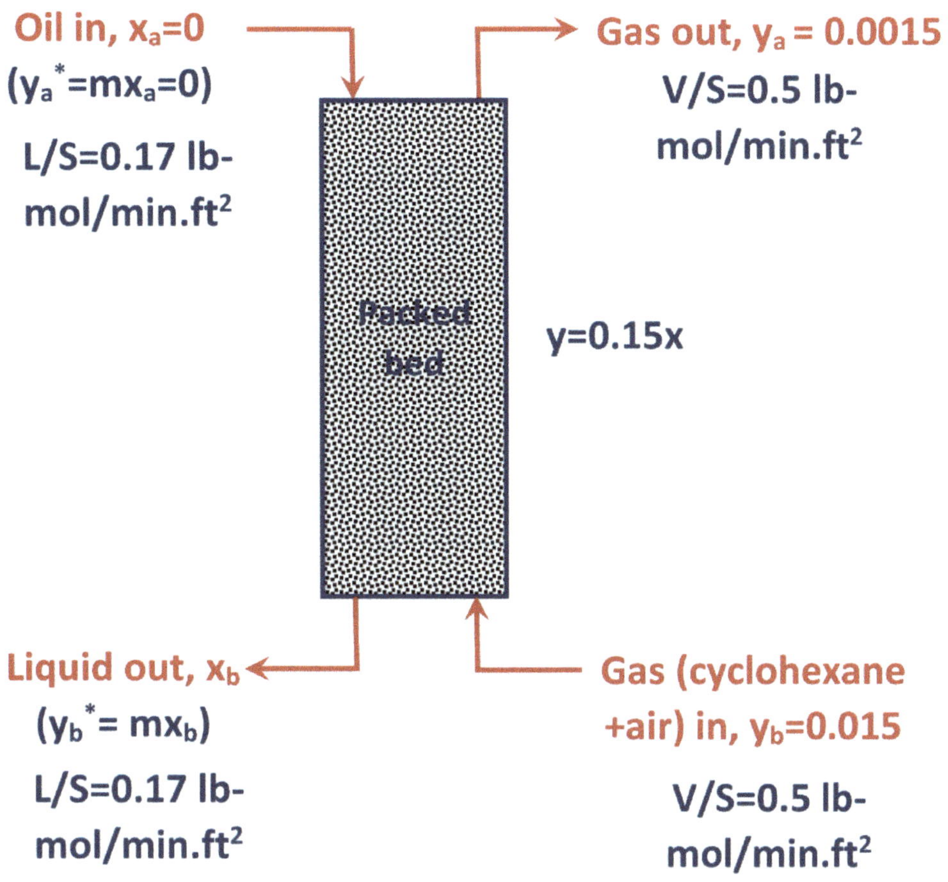

We first need to calculate x_b. From Equation (16.27):

$$y_a = y_b + (L/V)(x_a - x_b)$$

Upon rearrangement,

$$x_b = x_a + (y_b - y_a)/(L/V) = 0 + (0.015 - 0.0015)/(0.17/0.50) = 0.0397$$

Note that y_a is 10% of y_b , that is, $y_a = 0.10 \times 0.015 = 0.0015$

Thus all the mole fractions at inlet and outlet of the tower are now known.

a) the number of transfer units, NTUs

For linear operating and equilibrium lines with different slopes, we can use Equation (16.28) to determine the NTUs:

$$NTU = \frac{y_b - y_a}{(y - y^*)_{log-mean}} = \frac{0.015 - 0.0015}{(y - y^*)_{log-mean}}$$

$$(y - y^*)_{log-mean} = \frac{(y_b - y_b^*) - (y_a - y_a^*)}{ln\left(\frac{y_b - y_b^*}{y_a - y_a^*}\right)} = \frac{(0.015 - 5.955 \times 10^{-3}) - (0.0015 - 0)}{ln\left(\frac{0.015 - 5.955 \times 10^{-3}}{0.0015 - 0}\right)}$$
$$= 4.1993 \times 10^{-3}$$

$$NTU = \frac{0.015 - 0.0015}{4.1993 \times 10^{-3}} = 3.215$$

Note that $y_a^* = 0.15x_a = 0$ and $y_b^* = 0.15x_b = 0.15(0.0397) = 5.955 \times 10^{-3}$

b) the height of the transfer unit, HTU

From Equation (16.24):

$$HTU = \left(\frac{V/S}{K_y a}\right)$$

We need to estimate $K_y a$ to calculate HTU:

$$\frac{1}{K_y a} = \frac{1}{k_y a} + \frac{m}{k_x a} = \frac{1}{14} + \frac{0.15}{0.30}$$

$$K_y a = 1.75 \text{ lb-mol/h.ft}^3$$

$$HTU = \left(\frac{V/S}{K_y a}\right) = \frac{0.5 \times 60}{1.75} = 17.143 \text{ ft}$$

c) the height of the packed tower

$$h = (HTU) \times (NTU) = 17.143 \times 3.215 = 55.1\ ft$$

Problem 3

A packed tower is to be designed to absorb ammonia from an air stream using pure water as a solvent. The gas enters the tower at a mass flow rate of 1500 lb_m/h. The mass flow rates of liquid is 3000 lb_m/h. The tower is packed with 1.5-inch metal Pall rings.
Estimate the diameter of the tower such that the gas mass velocity is 70% of the flooding limit.
The properties of the fluids are: $\rho_V = 0.062\ lb_m/ft^3$, $\rho_L = 62.4\ lb_m/ft^3$, $\mu_L = 1$ cP
For 1.5-inch metal Pall rings, the packing factor $F_p = 32\ ft^{-1}$.

Solution

Refer to the schematic diagram of the packed tower shown below.

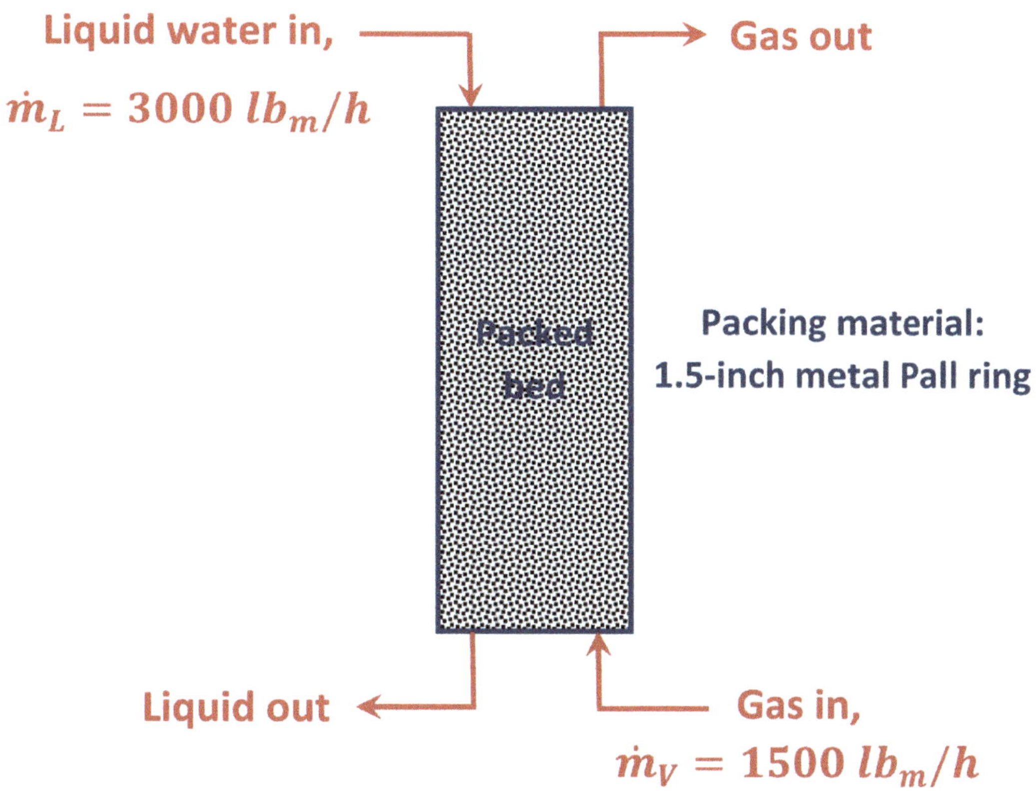

From Equation (16.42):

$$F_{LV} = \frac{\dot{m}_L}{\dot{m}_V}\sqrt{\frac{\rho_V}{\rho_L}} = \frac{3000}{1500} \times \sqrt{\frac{0.062}{62.4}} = 0.06304$$

Using Equation (16.43), we can estimate the flooding velocity:

$$log_{10}(G_V^2\,\psi) = -1.6678 - 1.085log_{10}(F_{LV}) - 0.29655[log_{10}(F_{LV})]^2$$

$$log_{10}(G_V^2\,\psi) = -1.6678 - 1.085log_{10}(0.06304) - 0.29655[log_{10}(0.06304)]^2 = -0.7927$$

$$G_V^2\,\psi = 0.161$$

From Equation (16.41):

$$\psi = \frac{F_p \phi \mu_L^{0.2}}{\rho_V \rho_L g_c} = \frac{32 \times 1 \times 1^{0.2}}{0.062 \times 62.4 \times 32.2} = 0.2569$$

Thus, the flooding mass velocity G_V is:

$$G_{V,flooding} = \sqrt{\frac{0.161}{0.2569}} = 0.7916\ \frac{lb_m}{s.ft^2}$$

Using Equation (16.49),

$$D = \sqrt{\frac{4\dot{m}_V}{\pi\alpha(G_V)_{flooding}}} = \sqrt{\frac{4 \times (1500/3600)}{\pi \times 0.70 \times 0.7916}} = 0.9785\ ft$$

Corrections in the Text

- Page 30, line 4: "form" → "from".
- Page 91, Problem 1, part (c): "Kremser equation" → "Kremser equation?".
- Page 91, Problem 2, part (a): "propane" → "ammonia".
- Page 91, Problem 2, part (c): "Kremser equation" → "Kremser equation?".
- Page 97, first line after Equation (6.1): "The full derivation of Clapeyron equation is given in the tutorial section" should be deleted.
- Page 153, Problem 1: "The feed mixture is 40 mole% benzene." should be "The feed mixture is 50 mole% benzene.".
- Page 201, Equation (9.70): "$RD + S$" should be "$RD - S$".
- Page 224, Problem 3: "The distillate product contains 98 mol% ethanol.." should be "The distillate product contains 78 mol% ethanol..".
- Page 256, Problem 3 solution: The following equation

  $$D = 2\sqrt{\frac{\dot{m}_V}{0.80\pi U_f\left(1-\frac{A_d}{A}\right)\rho_V}} = 2\sqrt{\frac{19000}{0.80\pi(1.28)(1-0.1)(3.5)}} = 1.291\ m$$

  should be replaced by

  $$D = 2\sqrt{\frac{\dot{m}_V}{0.80\pi U_f\left(1-\frac{A_d}{A}\right)\rho_V}} = 2\sqrt{\frac{(19000/3600)}{\pi(1.28)(1-0.1)(3.5)}} = 1.291\ m$$
- Page 307, Problem 3: "The relative humidity of benzene .." should be replaced by "The relative volatility of benzene ..".
- Page 392, solution to problem 2: The following

  $$V_1 = 29.37\ kg/h,\quad y_1 = 0.423$$

  should be replaced by

  $$V_1 = 529.37\ kg/h,\quad y_1 = 0.423$$

www.ingramcontent.com/pod-product-compliance
Lightning Source LLC
LaVergne TN
LVHW070854210926
848482LV00016B/937